T0326576

Renewables

A review of sustainable energy supply options

Renewables

A review of sustainable energy supply options

David Elliott

David Elliott is Emeritus Professor of Technology Policy at the Open University, where he has focused on renewable energy policy.

IOP Publishing, Bristol, UK

ISBN 978-0-750-31040-6 (ebook)
ISBN 978-0-750-31041-3 (print)

DOI 10.1088/978-0-750-31040-6

Version: 20130910

British Library Cataloguing-in-Publication Data
A catalogue record for this book is available from the British Library.

Published by IOP Publishing, wholly owned by The Institute of Physics, London

IOP Publishing, Temple Circus, Temple Way, Bristol BS1 6HG, UK

US Office: IOP Publishing, The Public Ledger Building, Suite 929, 150 South Independence Mall West, Philadelphia, PA 19106, USA

Cover image: Complex Systems, Dorset. © Stanislav Shmelev

Contents

Some common technical abbreviations used:

AC	Alternating current
AD	Anaerobic digestion
CCS	Carbon capture and storage
CCGT	Combined cycle gas turbine
CHP	Combined heat and power
CO_2	Carbon dioxide
CSP	Concentrating solar power
CPV	Concentrating photo-voltaics
HVDC	High voltage direct current
OWC	Oscillating water column
PV	Photo-voltaic (solar)
SRC	Short rotation coppice

Some other abbreviations:

DECC	UK Department of Energy and Climate Change
EU	European Union
FiT	Feed in tariff
NGO	Non-governmental organisation
MENA	Middle East North Africa

Acknowledgments and dedications

Thanks to David Finney, Godfrey Boyle, Keith Barnham and Tam Dougan for commenting on drafts or sections. Thanks also to the final year/Masters students at Bath, Cranfield, Exeter (UEC), Kingston, Loughborough, Oxford, Southampton and Warwick, with whom I tested out some of the ideas over the past few years. And of course to the OU students, on various related courses: this book is for you especially, and also for my son Oliver and the new generation of students, who I hope will see some of these ideas become reality.

Note on references. Where relevant and helpful (for extra information), throughout this book I have provided references to my original *Renew Your Energy* blogs. All the web links in the reference sections for each chapter were accessed in April 2013. The Department of Energy and Climate Change changed its web address mid-way through the writing of this book: in my references I used the new versions, although in some cases you will be automatically redirected to the new location www.gov.uk/government/organisations/department-of-energy-climate-change.

Author biography

David Elliott

David Elliott trained as a physicist and worked initially with the UK Atomic Energy Authority at Harwell and the Central Electricity Generating Board, before moving to the Open University, where he is now an Emeritus Professor. Whilst at the Open University, he was the co-director of the Energy and Environment Research Unit and Professor of Technology Policy in the Faculty of Mathematics, Computing and Technology. During his time at the Open University he created several courses in Design and Innovation, with special emphasis on how the innovation development process can be directed towards sustainable technologies. Professor Elliott has published numerous books, reports and papers, especially in the area of development of sustainable and renewable energy technologies and systems. Still very active in research and policy since his retirement, he also writes the popular blog *Renew Your Energy* on http://environmentalresearchweb.org/.

IOP Publishing

Renewables

A review of sustainable energy supply options

David Elliott

Chapter 1

Introduction

Renewable energy: an overview of the issues and options

Renewable sources of energy, sometimes simply called 'renewables', are increasingly being used to meet our needs. This book attempts to review the state of play and explain how and why this expansion can and should continue, and indeed accelerate.

1.1 Why renewables?

The planet's natural flows of energy, driven mainly by incoming solar energy, the climate and the hydrological system, are *renewable* sources of energy. They are naturally replenished and will not be exhausted if the flows are tapped to power machines or energy conversion devices. In that sense they are *sustainable* and can be relied on into the future. Most are also, arguably, sustainable in the sense of not resulting in environmental or social impacts which would preclude their continued use into the future.

Renewable energy sources were used historically to provide mechanical power for milling and also during the initial phase of the industrial revolution. Why would we want to return to their use now? Here are some key reasons:

- fossil fuels will not last for ever: we face peaks in oil production and also for gas and eventually coal, and costs will rise as demand increases and supplies are diminished;
- we cannot use all the fossil reserves we have without seriously disrupting the climate;
- nuclear fission has many problems: cost, safety, security, long-term fuel availability.

It may be that there are temporary solutions to some of these problems (e.g. carbon capture and storage, fast breeder nuclear reactors) but, costs for the moment apart, renewable sources have none of these problems. They are the only long-term option for energy supply we have at present. The global renewable resource potential is very large.

doi:10.1088/978-0-750-31040-6ch1

The power available can be presented in terawatts (TW: see *Units*, box 1.1 below). Incident solar radiation inputs around 90 000 TW each year to the planet, on a continuous basis.

Of this, around 1000 TW might be accessible in the form of solar energy directly, and 10 TW of wind, as well as wave, hydro, and (non-solar) tidal and geothermal energy (Jackson 1992). For comparison, globally there is about 18 TW of primary power demand. So there is much more available from renewables than humanity will ever need, if sufficient useable energy can be extracted at reasonable cost.

At present about 17% of total global energy is obtained from renewable sources, mostly traditional biomass and hydro, but the so-called new renewables, notably wind and solar, are expanding rapidly. By the end of 2012 there was 282 GW of wind capacity in place globally, 245 GW of solar thermal heating and 100 GW of solar photo-voltaics (PV). Rapid expansion continues, despite the global recession, with wind capacity expected to double over the next five years and PV perhaps treble, as it heads for grid price parity.

There have been some dramatic longer term projections of potential expansion. In 2008, the German *Energy Watch* group produced a 'high renewables' scenario, with

Box 1.1. Units and terms

'Power' is often confused with 'energy', and is also often used to mean 'electricity', whereas the latter is only one form of energy; heat and transport fuels are others.

The **power** of an energy device is measured in **watts** (W) and multiples of watts: kilowatts (1 kW = 1000 W), megawatts (1 MW = 1000 kW), gigawatts (1 GW = 1000 MW) and terawatts (1 TW = 1000 GW). A large wind turbine might have a full power rating or 'generating capacity' of 5 MW; the figure for a large conventional power station might be 1 GW; for a kettle, 1–2 kW.

Although wind turbines, nuclear power plants and so on are often called energy *generating* devices, strictly they are energy *conversion* devices, converting one form of energy into another, with inevitably some losses. Thermal energy generating capacity is differentiated from electricity capacity by adding 'th' or 'e' respectively (in brackets) after the power figure, e.g. MW(e). Note that the rated power capacity figure only indicates what the device could produce if it were able to work at full power output. In practice it will usually not be able to do that, for example given the variability of renewable sources like wind and solar, but also due to breakdowns, maintenance periods and so on. So, as I will be illustrating, the actual available generating capacity will be some fraction of the so-called 'nameplate' power rating capacity: see table 2.2 in chapter 2 for some typical *load factors*, reflecting the amount of energy that devices can actually deliver.

The amount of **energy** a device generates or uses (that is, converts from one form to another) depends on its power level and on the length of time it is operating at that level, so it is calculated as watts × time. The units of energy are kilowatt hours (kWh), megawatt-hours (MWh), and so on in 1000 Wh multiples (GWh and TWh).

The kilowatt hour is the unit used for selling electricity, so it should be familiar. I have (mostly) avoided the less familiar joule (J), though exajoules, 10^{18} joules, are used for some large quantities. To convert: watts = joules s^{-1}.

4450 GW of (non-hydro) renewables globally by 2030, meeting 30% of final total global energy demand, and 62% of global electricity needs (EWG 2008).

In 2011 WWF/Ecofys produced a scenario with renewables supplying nearly 100% of all energy globally by 2050 (WWF 2011). Stanford University Professor Mark Jacobson and Dr Mark Delucchi from UC Davis had come to similar conclusions in an article for *Scientific American* in 2009, later reworked as a paper in *Energy Policy*. It concluded that 100% of all electricity and possibly energy could be obtained globally from renewables by 2050 (Delucchi and Jacobson 2011, Jacobson and Delucchi 2009, 2011). More detailed regional studies have also emerged. Reports from the European Renewable Energy Council and the European Climate Foundation included scenarios with around 100% of the EU's electricity coming from renewables by 2050 (EREC and ECF 2010). In addition to the reports from trade/lobby groups, independent consultants Price Waterhouse Coopers produced a scenario of 100% electricity by 2050, covering Europe and North Africa (PWC 2011).

Getting to 100% globally or even regionally by 2050 may be challenging, but the usually cautious International Energy Agency has published scenarios with renewables supplying 75% of electricity globally by 2050 (IEA 2010), while the UN-backed Intergovernmental Panel on Climate Change has said that up to 77% of *all energy* could be supplied by renewables globally by 2050 (IPCC 2011).

The most recent review, the Global Energy Assessment, produced by an international team led by the International Institute for Applied Systems Analysis, noted that 'The share of renewable energy in global primary energy could increase from the current 17% to between 30% and 75%, and in some regions exceed 90%, by 2050' (GEA 2012).

Progress is being made. For example, the current EU plan is to get to 20% of total energy from renewables by 2020, which looks likely (though not certain) to be achieved. As a follow on, the EU 2050 Roadmap included a proposal for a target of between 55% in the lowest scenario and 75% in the highest with, in the latter case, 97% of electricity being supplied by renewables by 2050 (EC 2011). Similar growth is occurring elsewhere around the world, notably in China, which plans to get 15% of its energy from non-fossil sources by 2020, and may well do better than that. It is already the world wind leader.

Getting to 100% of all energy from renewables by 2050, globally, may be too much to expect, but some parts of the world could, and most countries should be able to move towards obtaining close to 100% of electricity from renewables by then, should they so wish.

1.2 Which sources are emerging?

Table 1.1 illustrates the basic planetary energy inputs that are available, but the practical problem is to develop technologies that can capture some of this energy efficiently. This book attempts to review the scene, looking at progress around the world, and also at the problems. As a brief initial overview, it may be helpful to go through the technology options, starting with those that are well established.

Hydroelectric plants use well-developed large-scale technology; they already supply around 17% of global electricity, and that is likely to grow. Geothermal energy from aquifers is widely used for heating, although on a much smaller scale, but geothermal

Table 1.1. Energy flows and sources (annual amounts in exajoules (EJ))

Total incoming energy	5.4 million
Available for uses	3.8 million
Solar radiation (heating air, land and oceans)	2 650 000
Hydrological cycle (rain feeding rivers)	1 080 000
Wind convection (and hence also waves)	11 700
Photosynthesis (in biomass)	1 260
Geothermal (conduction from hot rocks)	1 008
Ocean tides (gravitational)	94
For comparison: global energy use at present	502 (462 EJ hydro/biomass excluded)
Exajoules (EJ) = 1 billion billion joules (10^{18} joules)	

Data source: Boyle (2012).

heat is also used increasingly for electricity production. There is over 11 GW of electricity generation globally, with several new deep geothermal projects planned.

Solar heat collectors are now a very familiar sight around the world, with simple radiator-like units mounted on rooftops which, depending on location, can cut annual water/space heating fuel bills by 50% or more. Large-scale concentrated solar power (CSP) plants, focusing the Sun's heat to raise steam to drive electrical generators, have been developed for use in desert areas. The use of photovoltaic solar cells for electricity generation is now spreading widely around the world and, although they are still expensive, prices are falling rapidly as new cells emerge and mass production and markets grow.

Perhaps the most visible (literally!) new option is wind, harvested both on land and offshore, but wave and tidal energy technologies are also being developed rapidly. And, finally, new types of biomass use are emerging for electricity, heat and fuel production.

The above are all energy supply technologies, supplying electricity, heat or vehicle fuels. A sustainable energy future will also require attention to how this precious energy is used, to avoid wasting it. This book will not deal with energy conservation and efficiency (much less transport issues) in detail, important though they are, apart from where demand-side/end-use issues interact with supply-side issues.

It is clear that the potential for avoiding energy wastage is large, in part since most countries have, until recently, had access to relatively cheap fossil fuels and so little incentive to use them efficiently. Nevertheless, whatever is done to improve the efficiency of energy use—and we should do a lot—there will still be a need for energy sources. And renewables offer a range of climate-friendly low- or zero-carbon options.

1.3 What are the problems?

Some of the renewable sources are variable, wind and solar especially. It is therefore sometimes argued that they cannot be relied on to supply firm, continuous grid power. In fact, grid systems already balance the variation in availability of conventional plant output (including unexpected shutdowns) and the large daily cycles in demand. They use fossil plants, which run up and down to full power regularly, usually twice daily.

The smaller, slower variations in renewable outputs are easily balanced. The fossil plants simply have to cycle between high and low output a few times more often. Up to around a 20–30% contribution from variable renewables represents a small operational issue, adding slightly to the system costs, but in effect reducing the overall cost and carbon emissions as a result of not having to buy and use so much fossil fuel, even if only by a few percent.

However, if variable renewables supply a larger proportion of grid power, other balancing measures will be needed. One option is to exchange electricity between countries to balance local/regional variations, using long-distance, very efficient high voltage direct current (HVDC) supergrids (Czich 2011). In addition, demand can be managed using interactive load management to delay/shift demand when supplies are low, for example via smart grid systems that can switch off some non-critical energy loads temporarily.

Another option is to use some of the excess electricity produced by wind and other renewables when demand is low to generate hydrogen gas, store it and use it to generate energy when wind is low and demand is high. In addition, other forms of storage are possible; indeed, pumped hydro storage is already used to help balance variable wind energy output. Storable biomass can also be used for backup. As I will explain, there are many other options, including compressed air, liquid air and heat storage.

A more fundamental problem is that most renewable energy flows and sources are by their nature diffuse, so capturing significant amounts of energy often involves large land areas, most notably in the case of biomass grown for energy crops. Wind farms can also cover quite large areas, and although most of the land surrounding the turbine bases can still be used for farming, some people object to the visual intrusion of wind machines. Offshore location avoids both of these problems but, like on land location, may have an impact on wildlife. Solar on rooftops involves no additional land use, but there can be land-use conflicts in relation to ground-mounted solar farms, or large focused solar CSP plants. Large hydro can also have major land-use, social and environmental impacts.

Those and some other examples apart, the local impacts of renewables are, as I will be showing, usually low and very much less than the global and local impacts that follow from the use of fossil fuel based energy technologies, most obviously in terms of adding to climate change through the release of carbon dioxide gas into the atmosphere. Some argue that the risks and impacts associated with the use of renewables like wind and solar are also much less than those associated with nuclear power.

One final problem area is of course cost. Most of the technologies are relatively new, but costs are falling, with some now being competitive with conventional sources in some locations. I will look at the economics as I review each source and compare them later. However, one overall view is that, although the capital cost of deploying some renewables may initially be high, as these options develop and replace increasingly expensive and scarce conventional fuels, the costs of energy will reduce from what they would otherwise have been. After all, with most renewables, there are no fuel costs.

1.4 The structure of this book

Rather than using a classification by primary energy source (not very helpful since, tidal and geothermal apart, they are all solar driven), this book reviews the renewable energy options by basic physical and technological engineering type. It starts with fluid-driven mechanical power (wind, wave, tidal, hydro), then moves on to heat-based systems (solar thermal, biomass, geothermal) and finally light conversion (PV solar).

This 'power', 'heat' and 'light' classification system in some ways maps onto how renewable energy generation has been dealt with historically, firstly with rotational devices, then with thermodynamic systems and finally with direct conversion.

Although some useful parallels emerge, it is not a perfect classification. Since the industrial revolution, most rotational devices (turbines) have been driven by heat sources to generate electricity, and some modern renewable heat sources (e.g. some biomass, solar and geothermal) likewise drive rotational devices. However, wind, wave, hydro and tidal generators do not use heat, while solar heat collectors do not use rotating devices, and PV uses neither.

The three chapters on specific renewables are followed by a chapter on integration issues, including storage and grid balancing, and then a concluding chapter looks at national progress and the way ahead, including support systems and policy issues, with a short afterword looking at 'contrarian' views.

This is not a textbook. There are plenty of these available (e.g. Twidell 2006, Coley 2008, Harvey 2010, Boyle 2012). Instead the aim is to review what is happening around the world, so as to convey the sense of excitement that abounds in this new, rapidly expanding area of technological development, but also to look at the problems, including, crucially, environmental impact issues. After all, an attraction of renewables is meant to be their low impact. In what follows I have drawn extensively on material from my weekly *Renew Your Energy* blog, produced for IOP Publishing's Environmental Research Website. That will continue to provide updates should you need them: see box 6.4.

At the end of each chapter I have provided brief **summary points**. These are often simple, very tentative and general. Before making any form of judgment on the viability of any specific option or set of options, it is wise to 'do the numbers' and take a detailed quantitative look at efficiencies, costs, reliability and any hard engineering, financial or environmental data that are available.

Professor David McKay's generally excellent self-published book, *Sustainable Energy Without the Hot Air* (MacKay 2007), attempts to assess the various options in general terms, and this approach offers valuable awareness of some the physical limits of renewables, including land area. However, perhaps inevitably, there are also limits to this approach due, for example, to the need to simplify, aggregate and average out the data.

Although it does make what can be rather abstract data more understandable, I am not sure it is very helpful to derive and use figures for individual energy use, or averaged figures across whole countries. It is hard to avoid the latter (I have not managed to), but the data can disguise what may be very different patterns of energy use by various groups in a society and in different parts of a country. Moreover, practical engineering reality may avoid what seem like physical limits. For example, wind farms may be thought of as taking up a lot of room, but in reality, the only area lost to farming is that of the turbine tower bases and any access roads. That type of issue is one I have tried to address.

In this book I have tried to provide quantitative data to help readers make their own assessments of the options, but since this is a review of the field, I have also set these data in a wider policy and engineering development context. The final (main) chapter includes an invitation to test possible scenarios for the future based on renewables by using the Pathways 2050 model that has been developed by the UK Department of Energy and Climate Change (DECC), under Professor MacKay's guidance.

Having just made use of Pathways 2050 myself (for a Pugwash report), I can confirm that it provides a good way to focus attention on the numbers, but also alerts you to their shortcomings, both in technical terms and in terms of policy vision (Pugwash 2013). So while I think that it is vital to approach the energy future with hard-headed analysis, we also need vision to provide the momentum to resist current dogma and challenge the dominance of the status quo. The need for change is clear. The drive toward renewables is, I believe, part of that. The Chinese philosopher Lao Tzu said 'When storms come, some build walls, some are thrown by the wind, others build windmills'. I hope this book 'renews your energy' to play a part.

References

Boyle G (ed) 2012 *Renewable Energy* 3rd edn (Oxford: Oxford University Press)

Coley D 2008 *Energy and Climate Change: Creating a Sustainable Future* (Chichester: Wiley)

Czich G 2011 *Scenarios for a Future Electricity Supply* (London: IET)

Delucchi M A and Jacobson M Z 2011 Providing all global energy with wind, water, and solar power, part II: reliability, system and transmission costs, and policies *Energy Policy* **39** (3) 1170–90

EC 2011 Energy Roadmap 2050: a secure, competitive and low-carbon energy sector is possible, European Commission, Brussels, http://europa.eu/rapid/pressReleasesAction.do?reference=IP/11/1543&type=HTMLIEAE

ECF 2010 Roadmap 2050, European Climate Foundation, Brussels, http://www.roadmap2050.eu

EREC 2010 Rethinking 2050, European Renewable Energy Council, Brussels, http://www.rethinking2050.eu

EWG 2008 Renewable Energy Outlook 2030, Energy Watch Group, Berlin, http://www.energywatchgroup.org/Renewables.52+M5d637b1e38d.0.html

GEA 2012 Global Energy Assessment, International Institute for Applied Systems Analysis, Austria, http://www.iiasa.ac.at/web/home/research/researchPrograms/Energy/Home-GEA.en.html

Harvey D 2010 *Carbon-Free Energy Supply* (London: Earthscan)

IEA 2010 Energy Technology Perspectives: Scenarios & Strategies to 2050, International Energy Agency, Paris, http://www.iea.org/techno/etp/etp10/English.pdf

IPCC 2011 Special Report on Renewable Energy Sources and Climate Change Mitigation, Intergovernmental Panel on Climate Change, Switzerland, http://www.srren.org

Jackson T 1992 Renewable energy: summary paper for the renewable series *Energy Policy* **20** (9) 861–83

Jacobson M Z and Delucchi M A 2009 A plan to power 100 percent of the planet with renewables *Scientific American* http://www.scientificamerican.com/article.cfm?id=a-path-to-sustainable-energy-by-2030

Jacobson M Z and Delucchi M A 2011 Providing all global energy with wind, water, and solar power part I: technologies, energy resources, quantities and areas of infrastructure, and materials *Energy Policy* **39** (3) 1154–69

MacKay D 2007 *Sustainable Energy Without the Hot Air*, available as an online book, http://www.withouthotair.com/

PWC 2010 A roadmap to 2050 for Europe and North Africa, Price Waterhouse Coopers, London, http://www.pwc.co.uk/eng/publications/100_percent_renewable_electricity.html

Pugwash 2013 Pathways to 2050: three possible UK energy strategies, British Pugwash, London, http://www.britishpugwash.org/recent_pubs.htm

Twidell J and Weir T 2006 *Renewable Energy Resources* (Abingdon: Taylor and Francis)

WWF 2011 The Energy Report—100% renewable energy by 2050, World Wide Fund for Nature with Ecofys, London, http://www.wwf.org.uk/research_centre/research_centre_results.cfm?uNewsID=4565

Chapter 2

Power

Power for machines: hydro, wind, wave and tidal power plants

Historically, humankind initially relied on human or animal muscle power to move things and drive simple machines, like corn grinding mills and water pumps for irrigation. That is still the case in some parts of the world. Muscle power derives from food. Depending on how that is sourced, it may not be sustainable, i.e. it may not be a naturally renewed resource. However, corn grinding mills and the like were later driven by mechanical power captured by rotating devices located directly in natural energy flows, of wind or water, which are sustainable resources. We have since moved on from water wheels, tidal mills and windmills to develop more sophisticated machines for generating electricity.

2.1 Hydro

Hydro plants were the first large renewable electric power systems to be developed. Hydroelectric power plants trap water behind dams on rivers, large and small, to create a head of water (i.e. potential energy) which can be released to produce kinetic energy to drive high-speed turbines for electric power generation. They can produce large amounts of energy on a continuous, reliable basis; the overall annual load factor (energy out as a percentage of the potential maximum energy) for hydro projects globally is around 44%, but it is location dependent. For some sites it is much higher.

Hydro plants are the largest existing renewable sources of electricity, with around 874 GW of generating capacity installed worldwide. The technology is mature, and deployment is still spreading. Hydro provides almost all of the electricity on the grids of many developing countries, for example (in 2008) nearly 100% in Albania, Angola, Bhutan, Burundi, Costa Rica, DR Congo, Lesotho, Mozambique, Nepal, Paraguay, Tajikistan and Zambia, as well as 60–90% in 30 other developing countries. In addition it provides nearly all the electricity in Norway, most of it in Iceland, and around 60% in Austria, Canada, New Zealand and Sweden (Lenz 2012).

Since the energy output is dependent on the head height, sites which can accommodate taller dams can produce more energy, and this follows a square law. Doubling

doi:10.1088/978-0-750-31040-6ch2

the head gives four times more energy on average. So in energy terms at least, the bigger the better. Moreover, trapping a large mass of water in the reservoir behind the dam will help to give a guaranteed supply, although there will be cost trade-off and site limitations.

The energy source is ultimately solar heat, which drives the hydrological cycle, but since that is climate and weather related, the energy resource at any particular location and time can vary. Indeed, with decreased rainfall in some areas in recent years, output from some hydro plants has fallen. For example, load factors have fallen to 22% in poor years in the UK, although 38% is quoted by the UK Department of Energy and climate change (DECC) as a national average (DECC 2012a). But in some parts of the world, droughts have meant that hydro outputs have become increasingly unreliable. This is likely to get worse with climate change.

The interactions between hydro projects and the environment are also two-way. Large projects can have significant environmental impacts. Many environmental/development organisations, including World Wide Fund for Nature (WWF), Friends of the Earth (FoE) and Oxfam, while backing smaller-scale hydro, have opposed large hydro projects because of the large social and environmental impacts. The social dislocation resulting from flooding areas for new reservoirs is an obvious issue, but there are also wider ecological issues. For example, the World Commission on Dams non-governmental organisation has claimed that, in some hot climates, biomass carried downstream can be collected by the dam and can rot, generating methane, so that net greenhouse emissions can be more than from a fossil plant of the same energy capacity. It is not just a matter of any initial biomass trapped when the hydro reservoir was first filled, but a continuous ongoing process of fresh biomass decay (WCD 2000, Harvey 2010).

The industry does not accept this. It says emissions are not a general problem and, in any case, there are remedial options (IHA 2002). While some critics still insist that hydro may not be quite such an attractive renewable source as often suggested, there remains a strong push for more hydro, and large schemes. For example, the African Union, World Energy Council, International Commission on Large Dams, International Hydropower Association and others have joined together to insist that hydro is an important solution to some of Africa's problems. They note that 'During the past century, hydropower has made an important contribution to development, as shown in the experience of developed countries, where most hydropower potential has been harnessed. In some developing countries, hydropower has contributed to poverty reduction and economic growth through regional development and to expansion of industry. In this regard, we note that two-thirds of economically viable hydropower potential is yet to be tapped and 90% of this potential is still available in developing countries. In Africa, less than 7% of hydropower potential has been developed'.

However, they recognise the need to take the environmental issues seriously: 'We firmly believe that there is a need to develop hydropower that is economically, socially, and environmentally sustainable' but say that 'a number of lessons have been learnt from past experience. Governments, financing agencies and industry have developed policies, frameworks and guidelines for evaluation and mitigation of environmental and social impacts, and for addressing the concerns of vulnerable communities affected by hydropower development'. They note that 'the key ingredients for successful resettlement include minimization of resettlement, commitment to the objectives of the resettlement by the developer, rigorous resettlement planning with full participation of affected communities, giving particular attention to vulnerable communities'. They add 'The decision-making process should incorporate the informed participation of the vulnerable communities and those negatively affected, who must in all circumstances derive sustainable benefits from the project. The costs of social and environmental mitigation measures and plans should be fully assessed and integrated in the total cost of the project' (AU *et al* 2009).

The overall conclusion, though, remains that large hydro is beneficial. They point to giant potential projects like the proposed £40bn Grand Inga on the Congo river. Its 40 000 MW capacity could, supporters say, generate more than 280 TWh $year^{-1}$ of electricity at less than $0.01/kWh. For comparison, diesel generators, widely used in Africa, cost from $0.15/kWh to $0.30/kWh. It could, backers say, double the amount of electricity available on the continent and jump-start industrial development, bringing electricity to hundreds of millions of people as well as exporting electricity to South Africa, Nigeria and Egypt, and even Europe. It would supply twice as much electricity as the world's current largest dam, the Three Gorges in China, which is 18.2 GW (though soon to be 22.4 GW).

However, not everyone is so keen. Terri Hathaway, Africa campaigner with International Rivers, a watchdog group monitoring the Grand Inga project, said that 'As it stands, the project's electricity won't reach even a fraction of the continent's 500 million people not yet connected to the grid. Building a distribution network that would actually light up Africa would increase the project's cost exponentially. It would be very different if rural energy received the kind of commitment and attention now being lavished on Inga' (Hathaway 2008).

Grand Inga was proposed in the 1980s but never got beyond feasibility studies because of political turmoil in central Africa. Now there seem to be prospects for it to go ahead and be completed by 2022. One big change is that banks and private companies can earn high returns from the emerging global carbon offset market and, in some cases, from Clean Development Mechanism credits.

The debate on large hydro continues, with the industry presenting hydro as a fully sustainable option (IHA 2013). While it is clear that hydro has many attractions and that Africa and other places in the developing world need energy, there are also clearly counter-views about whether hydro, especially large hydro, is the best option. Large projects are expensive and involve big companies who, some fear, may not be concerned about local impacts.

Certainly there have been some bitter battles fought over some projects and large-scale hydro remains a politically contentious issue in many parts of the world. Quite apart from local impact issues, it is sometimes argued that large centralised projects may in any case be the wrong answer for Africa and other similar locations. The very large distances involved make it unlikely that grids could ever cover the entire continent. As with the Grand Inga project, much of the electricity seems likely to be exported on high voltage direct current (HVDC) links to remote markets, rather than be used locally. It is argued that local decentralised power may make more sense. These can be micro hydro, wind, biomass or solar technologies which can be installed quickly with low local impact and with a potential for direct local involvement, and also providing the possible creation of local manufacturing enterprises to build the equipment.

While some newly industrialising countries like Brazil, China and India, which already have large hydro capacities, are looking to further expansion, within most of the already developed world, there is much less potential remaining for large projects, although Russia, for example, still has plans for significant expansion. However, there may be a wider potential for smaller-scale projects. Small hydro contributes about 3% to the total electricity generation in Europe, with over 17 800 small hydro schemes and a total installed capacity of 12 333 MW in the EU-27. Moreover, the European Small Hydro-power Association considers that there are major opportunities for further small projects in the EU, as well as worldwide: 'the remaining European energy potential for small hydropower is still considerable; in particular the Balkan region offers a booming market with excellent opportunities for investment as it contains huge potential and offers very good conditions with predictable economic and political development'. Furthermore, outside of Europe, the Emissions Trading Scheme and Clean Development Mechanism markets 'bring new opportunities for the sector. In fact, approximately 90% of all clean development mechanism projects in China are small hydropower' (ESHA 2013).

Although it is site-specific, small hydro in general is usually seen as less economically viable than large projects, at least in generation cost terms, but small projects are faster to build and so easier to finance, and can offer local social and economic benefits.

It is claimed that they can also have lower environmental impacts. As I will be showing later, there are similar issues concerning the relative merits of large and small tidal barrage projects, which have many features in common with low-head hydro projects, e.g. they use similar turbine generators. However, whether small hydro (or tidal barrage) projects really do have less environmental impact *per kWh produced* than larger projects is not clear, given their lower operating heads and energy output/kW installed. It may depend on the site (Clarke 2012). Nevertheless some small 'run of the river' schemes (with no large dams or reservoirs) can be economically attractive (Williams 2012).

It is worth noting that that there are around 48 000 small hydro projects in China, and that globally there are around 85 GW of small hydro projects (under 10 MW each), so it is not a trivial resource. Indeed it is expanding (it has been said that only around 25% of the global potential has been used so far), with very small 'micro hydro' schemes also finding niches around the world. See box 2.1 for some UK examples of small projects.

Hydro schemes, both small and large (but more likely large), may get a boost from the need to provide balancing services to compensate for the variability of renewables like wind energy. Excess electricity from such sources can be used to pump water uphill behind hydro dams to be stored temporarily in their reservoirs, ready for generating electricity via the hydro turbines when needed. There is more than 127 GW of pumped-storage hydro capacity already operating around the world, and significant growth in the market is projected, in Europe especially (JRC 2013).

In the EU, some projects make use of existing hydro plants and their reservoirs, linking them up. For example, Alpine mountain ranges are being used to develop linked hydro projects with pumped storage capability. In the Voralberg area in Austria, a set of relatively small reservoirs at different levels in the Alps have been linked up, with several being modified so that surplus power from the nearby Bavarian wind turbines

Box 2.1. Small hydro in the UK

The UK only obtains relatively small amounts of its electricity from hydro, from around 1.5 GW installed, mostly in small- to medium-sized projects in Scotland and Wales, but a report for the Environment Agency identified almost 26 000 sites that were theoretically suitable for small hydro schemes which could take around 1178 MW of capacity, meeting about 1% of the total predicted 2020 electricity demand. However, not all the sites identified were practical, due to environmental and technical constraints. Even so, the report said there were nearly 4000 unused sites in England and Wales with the potential for generation without damaging the environment.

In parallel, one study estimated that there could be 1.2 GW of new hydro capacity in 7043 schemes in Scotland. The study took into account micro hydro schemes under 100 kW, with many being likely to obtain support under the UK's Feed-in Tariff for small projects (FREDS 2008).

New low-impact technologies, such as Archimedes screws, are claimed to offer advantages, e.g. for low-head or 'run of the river' use, and there are some interesting new projects (Halliday 2013).

can be used at off-peak periods to pump water back up to higher reservoirs (Schmoeller 2006).

Some projects involve radically new schemes. In the Harz Mountains in Germany, old mine workings are being looked at as possible sites for new underground pumped storage facilities. The UK already has one like this, the 1.8 GW Dinorwig plant in Wales.

There are some problems. For example, if the reservoirs are full (e.g. from snow melt), extra water cannot be added. Denmark does make use of electricity imported from hydro in Norway when its large wind capacity is not delivering enough power, and when Danish wind is high but national energy demand is low, it then exports any excess to Norway, to be stored or used directly. But Denmark gets charged more for the imports than it receives for the exports. It is a hydro suppliers' market, raising the cost of wind (Bach 2011). I will be looking in more detail at the role of pumped storage as a method of balancing the output from variable renewables in chapter 5.

Established large hydro delivers some of the cheapest electricity on the grid in many countries but, as I have noted, there are some environmental question marks. There are also some safety problems with large hydro. Dams can and do sometimes break, rapidly inundating areas beneath. For example, between 1970 and 1992, 4000 people were killed by hydro dam failures globally, and this is fairly typical, making hydro one of the worst energy options in terms of deaths per kWh although, if the health impacts of air pollution are added in, coal still leads, even more so if climate change impacts are included. I will be looking at the relative merits of the various renewable energy options in chapter 6.

2.2 Wind power

Wind is the second major form of renewable energy that has been used to produce electricity. Developing on the early multi-vane wind pumps used for irrigation, in recent times electricity-generating wind turbines have expanded from small kilowatt battery wind-chargers to multi-megawatt grid-linked devices, with increasing numbers offshore, in some cases in floating units.

In terms of turbine developers, manufacturers and equipment suppliers, Denmark (e.g. Vestas) and then Germany (e.g. Siemens) initially took the lead, followed later by the USA (GE) and then China (Sinovel). It has become a very large, rapidly expanding and competitive global market.

By the end of 2012, over 282 GW of wind generation capacity had been installed around the world, mostly on land, but with around 5.4 GW offshore. The 2012 installed capacity data, by country, are shown in table 2.1, along with some data on energy production in 2011 (the most recent available). The actual energy outputs will depend on location and wind speeds. Hence some countries achieve less, or more, from the same installed capacity.

Horizontal axis propeller-type devices remain the most popular design (though they are not really propellers, since they do not propel anything), but vertical axis devices have the attraction that they can operate regardless of the direction of the wind and have the generator at the base, making them more stable and perhaps better suited for offshore use. Wind turbines have moved upscale, with 2 MW units now being common on land and 5 MW units offshore, while 10 MW devices are under development for offshore use.

Winds are caused by the differential heating of air, land and sea by the Sun, modified by topographical features and the planet's rotation, so wind energy is basically indirect solar energy. The basic physics of wind flow across rotating blades means that, for propeller-type devices, the energy available is proportional to the square of the rotor

Table 2.1. Wind power generation capacity and energy outputs.

Wind capacity at end 2012 (GWEC 2013)			Wind energy outputs in 2011 (EIA 2013)	
Country	(MW)	% World total	Country	TWh
China*	75 564	26.8	USA	119.7
USA	60 007	21.1	China	73.2
Germany	31 332	11.1	Germany	46.5
Spain	22 796	8.1	Spain	42.4
India	18 421	6.5	India	26.0
UK	8 445	3.0	UK	15.5
Italy	8 144	2.9	France	12.2
France*	7 196	2.5	Italy	10.1
Canada	6 200	2.2		
Portugal	4 525	1.4	EU total	149.1
Rest of world	39 853	14.1		
Total	282 482	100	Total world	341.5

*Provisional; e.g. the Chinese State Electricity Regulatory Commission quoted 61 GW by end 2012, possibly since not all is grid linked. Also, some output is being curtailed due to lack of grid capacity, which may be one reason why the TWh figures are low for China (Qi 2013a, 2013b).

radius length and the cube of the wind speed. So, in physical terms, within the constraints imposed by blade material strength, the bigger the better, and since wind speeds are usually higher in hilly areas, the higher the better (Taylor 2012).

Some of the deployment problems associated with wind energy follow on from these basic physics-defined mechanical rules. Upland sites are usually environmentally sensitive, and large machines can be visually intrusive. That has led to significant opposition to some wind farms in the UK. To an extent it is an aesthetic issue; some people like the look of wind turbines and enjoy the idea of harvesting a free, clean resource. That said, sensitive location is essential to avoid resistance to projects.

Scale is obviously an issue. Large machines are much more intrusive than small devices, but also much more efficient. Depending on the location, domestic-scale 1–2 kW micro wind turbines may rarely be very useful. There are clear economies of scale due to the square and cube laws linking output to blade size and wind speed. So micro wind turbines will not be very efficient in most typically low wind speed urban areas. Just doubling the wind speed from 4 m s^{-1} (good urban site) to 8 m s^{-1} (good elevated rural site) would yield 8 times more wind potential. Indeed a UK study of micro wind devices found that 'in many urban areas they are unlikely to pay back either their [embodied] carbon emissions or the home owner's costs for installation and maintenance' (BRE 2007). By contrast, a wind farm in a windy site can have large MW machines which will generate far more energy per kW installed and per £ invested than the equivalent number of domestic 1 kW micro turbines.

That does not mean that medium-sized machines cannot find application for local community use. There are trade-offs between scale, cost and utility. A UK study, nicely entitled 'Power In Numbers', claimed that, while the cost of electricity from individual domestic-scale devices might be £3472/MWh in a village setting, with average wind speeds of 6 m s^{-1}, for a larger project serving 30 houses it might only cost £263/MWh. Moreover, for 100 houses it would be £164/MWh. It concluded: 'Where possible, communities should be encouraged to work together to deploy the largest possible turbines, as opposed to series of individual installations' (EST 2008).

Human concerns like visual intrusion apart, wind turbines seem to be relatively benign, although there have been concerns about the impact on birds and bats. There were some early problems with multiple bird strikes when wind farms were located in migratory paths, e.g. in Southern Spain and California. Clearly sites like that should be avoided and the incidence of casual impacts reduced to the minimum possible by sensitive location. However, birds tend to avoid moving objects, and collisions with power grid lines are usually much more of a threat. Observation of 500 m of power lines linked to a 400 MW conventional power plant in Spain estimated that they electrocuted 467 birds, and an additional 52 were killed in collisions with lines and towers over the course of two years (Sovacool 2009). This all has to be put in perspective. Although all accidental animal deaths should be avoided, cats kill vastly more birds than do wind turbines.

The UK Royal Society for the Protection of Birds has broadly supported well-sited wind farms, but issues remain concerning the impact on bats. A study of bat deaths at a local wind farm by the University of Calgary found that the majority of migratory bats there were killed because a sudden drop in air pressure near the blades caused injuries to the bats' lungs, known as barotrauma. Although the respiratory systems in birds can withstand such drops, the physiology of bats' lungs does not allow for the sudden change of pressure. TransAlta, a Canadian renewable energy company, looked at some possible solutions. They tested a revised operating procedure at the same site. Slowing turbine blades to nearly motionless in low-wind periods significantly reduced bat mortality. Professor Robert Barclay, who led the University of Calgary study, noted that 'biologically, this makes sense as bats are more likely to fly when wind speeds are relatively low'. The operational change was found to reduce bat deaths from wind turbines by up to 60% without significantly reducing the energy produced (Baerwald *et al* 2008).

Noise has also been claimed to be a problem in some locations. Some early machines were relatively noisy, but modern designs are very much less so. Visitors to wind farms are often surprised by how quiet they are, with just the blade swish sound, often masked by the sound of wind in nearby trees and bushes and certainly much less noticeable than the noise from any nearby road traffic. Modern machines are often designed so that rotation speed varies with the wind speed rather than, as with earlier machines where rotation was at a fixed speed, locked into step with the grid system. In the new variable speed devices, on-site electronic systems adjust the output to match grid requirements. It is worth doing, since then the turbine blade outer sections (which are the most energy productive part) can move at speeds similar to that of the wind. The result is better energy capture and less aerodynamic noise. The advent of gear-free transmission systems has also helped reduce gear train rumble, and acoustic baffles can reduce it further.

Nevertheless some people are sensitive to noise even at very low levels; for example, some people cannot sleep in a house with a refrigerator running, and sub-sonic noise (low frequency inaudible infra-sound) is claimed to be a problem for some people in some locations. The official view remains that there are no significant health impacts from wind farm noise but, to avoid nuisance, there are limits and separation rules imposed. However, those living nearby may feel differently: 'annoyance' is a subjective response (Haggett 2012).

The debate over noise can be heated, with local (and national) anti-wind campaigners adopting intransigent positions (Elliott 2009, 2010). The validity and analysis of some of the data have been disputed, given their reliance on small samples and anecdotal evidence (Donald 2012). That led an Australian academic to note wryly that 'pre-existing negative attitudes to wind farms are generally stronger predictors of annoyance than residential distance to the turbines or recorded levels of noise' (Plumer 2012).

While some wind farm developers may get a little exasperated at times, when faced with intransigent opposition, much attention has been given to reducing impacts via machine design and location, and in general the renewable energy community has been concerned to try to ensure that environmental and health issues are dealt with properly if they prove to be significant (Cummings 2012). After all, one of the reasons for adopting renewables like wind power is to reduce local and global environmental impacts. Some sites may be inappropriate, for example some un-degraded peatlands (Smith *et al* 2013), but in general it does seem clear that any local impact from wind projects will be small and will be vastly outweighed by the global (and hence also local) climate benefits.

Certainly, despite some local opposition and hostility from some conservation and preservation organisations, wind farms remain very popular with the public, even in the UK, where opposition has perhaps been highest. There seems to be something of a rural–urban split, perhaps reflecting city dwellers' perceptions of rural areas as leisure resources. In an Ipsos MORI poll in 2012, 62% of countryside dwellers said they found the visual impact of wind turbines acceptable, compared to 57% of the urban sample (Elliott 2012a). A 2012 opinion poll for DECC found that overall 66% supported on land wind, with 12% opposed, 4% strongly, while 73% of the sample backed offshore wind (DECC 2012b). Globally wind is even more popular. An Ipsos poll of 24 countries in 2011 found that 93% of those asked backed the use of wind energy (IPSOS 2011).

As well as being generally popular, it is clear that wind power is economically attractive. On land wind can be the cheapest of the large-scale new renewables and in some locations can supply electricity at prices competitive with those from conventional energy sources (Sourcewatch 2010). In the UK, a 2011 study by consultants Mott MacDonald (discussed further in chapter 5) found that the then current cost of electricity for on land wind was £83–90/MWh, compared to £96–98/MWh for nuclear and around £100/MWh for gas-fired plants, adjusted for the cost of carbon capture (Mott MacDonald 2011). In 2012, Cambridge Econometrics claimed that the UK economy would be £20bn a year better off by 2030 if it favoured offshore wind over gas-fired generation. This report, for WWF and Greenpeace, calculated that, as well as increasing gross domestic product by 0.8% by 2030, carbon emissions from the power sector would be two-thirds lower under the wind-based scenario, reducing the UK's carbon footprint by 13% (WWF 2012). With cost continuing to fall, projections like this may prove correct: the savings on fuel will mount.

2.2.1 Offshore wind

Offshore wind is currently more expensive than on land wind, due to the difficulty of installing and maintaining machines offshore, and the need to have expensive undersea grid links back to shore. New technology (including floating wind turbines) should be able to reduce the installation costs, and once there are multiple arrays in place, the cost per turbine of shared grid links will fall (UKERC 2010). The attractions of going offshore are very significant; offshore wind speeds are higher and more reliable since there are no topographical interactions, and that is where the very large, and much less constrained, resource is. For example, the UK resource has been estimated as more than enough to supply all the UK's electricity needs, while leaving plenty for export (PIRC 2010). So there is much effort being expended on reducing cost. Studies for the UK government suggest that it should be possible to reduce the cost of offshore wind in the UK by almost 30% from £140/MWh to £100/MWh by 2020, with the right actions from the industry.

With around 2.7 GW already installed offshore by 2012, there are some fascinating developments under way in the UK for future technology developments, supported by various government schemes. For example, 104 offshore wind turbine tower support designs competed for support under the Carbon Trust's Offshore Wind Accelerator programme. Four were short-listed for further development: Keystone's wonderfully named

'Inward Battered Guide Structure' (a 'twisted jacket' tripod tower design); SPT's Self-Installing wind turbine (it jacks itself up into position); Harland and Wolff's Universal Foundation design (a massive sink plunger-like bucket sitting on the seabed); and the Gifford BMT Freyssinet Gravity Base foundation (a huge mass sitting on the seabed).

More dramatic still are the various designs for floating devices. Some still have seabed supports, sometimes using tension leg designs with tethers to the seabed (an approach borrowed from the offshore oil industry) like the Dutch Blue H device (Blue H 2013) and Portugal's Windfloat (Windfloat 2012). However, fully floating spar buoy devices are being tested to allow location in deep water further out to sea, like the Norwegian Sway and Hywind projects, both of which have been tested at sea (Hywind 2012, Sway 2013).

All of the devices mentioned so far use conventional horizontal axis propeller-type turbines, but vertical axis designs are also being developed for use offshore. Examples of floating/tethered systems include the UK Nova project's V-shaped Aerogenerator X design (Arup 2010) and the French H-shaped Vertiwind (Gatto 2011). EU/RISO's Deep Wind project is an 'eggbeater' shaped Darrius design (RISO 2013).

It is clear that there is much activity in this area (Elliott 2012b, 4cOffshore 2013). It is predicted that the cumulative global market for floating wind turbines could exceed £200bn by 2030. See box 2.2 for more examples and grid linking developments in the EU.

The EU and the UK in particular have led with offshore wind. By the end of 2012, the UK had 1000 turbines in place and plans to have around 18 GW installed offshore by 2020, while Germany wants to have 10 GW by 2020 and France wants 6 GW by 2020.

Box 2.2. Floating offshore wind turbine arrays and supergrid networks

Although there are plans for offshore floating systems in the Atlantic (off the coasts of the EU and the US), and also in the Pacific (off Japan), one of the most intriguing floating offshore wind concepts is destined for the Mediterranean. The Swedish Hexicon design has a series of turbines mounted on a hexagon-shaped lattice platform, which can rotate around a central axis to align with wind direction The Maltese government plans an eventual 36-turbine scheme located 20 km off the island's north-east coast in water depths of 100–150 m, with a 460 m wide platform. Being further out to sea it would be much less visible than an inshore wind farm. The areas around the Maltese islands are considered to be too deep to allow for the economical and feasible construction of fixed monopole wind farms (Hexicon 2013).

There could be even more dramatic developments ahead, as more offshore projects emerge in the North Sea. The EU looks like being first to establish an offshore grid to link projects by HVDC undersea grids (Offshore Grid 2011). One intriguing idea is a huge circular supergrid ring mainly around the edge of the North Sea, with nodal links to wind farm arrays, as proposed by a Dutch design group (We@Sea 2010). But even the individual floating devices will be novel enough. Some are like fishermen's bobs/floats, tethered to the seabed, but angled in the water and able to tilt slightly, as with the Sway design (Sway 2013). Although a 1:6 scale Sway test model sank in bad weather off Norway in 2011, it is early days in this field, and the technology is developing rapidly, with Japan's Forward project, off Fukushima, adding a new sense of urgency (RT 2013).

There are many new projects, moving into deeper water. The largest EU project so far is the HiPRWind R&D programme, aiming to test a turbine on a triangular floating structure off the coast of Spain, 1.7 km from the Basque Country in 50 m deep waters (Hyperwind 2013).

The EU and the UK especially are fortunate in having relatively shallow water and non-sloping seabeds off some coasts, which is where the first projects have been sited, fixed to the seabed. However, floating systems should be cheaper, and as well as the EU projects just mentioned, floating devices are now being planned in countries without the EU/UK's geographical advantage, including the USA, China and Japan.

Japan is keen to press ahead with offshore wind projects, as it attempts to move away from reliance on nuclear power, following the 2011 Fukushima nuclear accident. Interestingly, all but one of its existing on land wind turbines survived the 2011 tsunami, including one on a coastal causeway. Fully offshore projects make a lot of sense in a crowded country where land is scarce. In 2013, a 2 MW floating wind turbine was installed 16 km off the coast from Fukushima, with two 7 MW units planned to follow (RT 2013).

That may lead to 80 more by 2020 in a 1 GW offshore wind farm, with more to follow elsewhere. Japan's potential offshore resource is perhaps 100 GW, with a medium-term target of at least 25 GW being called for (JWPA 2010).

China also has ambitions for offshore wind, which will play an increasing role in its huge wind programme. The Chinese Wind Power Development Roadmap 2050 sets a target of 200 GW installed capacity by 2020, 400 GW by 2030, and 1000 GW by 2050. It gives priority to on land wind before 2020, while experimenting with pilot near-shore projects, with 5 GW planned by 2015 and 30 GW by 2020. But from 2021 to 2030, on land and offshore wind will get equal attention, with experiments with pilot projects far offshore. From 2031 to 2050, it will support all-round development of on land (in the eastern, central and western regions) and far and near offshore (Qi 2011, PRNewswire 2013).

The US has a very large on land wind programme (over 60 GW so far), but it has been slower to get started on offshore wind, although projects are being proposed (Deepwater Wind 2012) and, in 2012, the US government announced $180m funding for four offshore wind projects to accelerate the sector. It is claimed that the USA could have 10 GW of offshore wind capacity by 2020, with 5 GW already planned. To link them, an underwater HVDC grid has been proposed off the east coast (Windconnection 2013).

Offshore projects obviously have to be carefully assessed in terms of impacts (CEFAS 2004), but studies have not found significant environmental issues. Crustaceans seem to like the foundations, although sea mammals stay clear, as do birds and fish (Lindeboomet 2011, Bayar 2013). There may be navigation issues, but offshore wind arrays, with warning lights and radio transponders installed, might actually help reduce hazards by marking out safe sea lanes.

2.2.2 Grid balancing and costs

Wind, whether on land or offshore, is a variable resource. I will be looking at that issue below, since it is common to several other similarly weather-dependent renewables. So far, with relatively low percentages of renewables on grids, it has not been a major issue. But it may be worth noting here that, since wind has been the first of the new variable sources to be developed, it has had to address this issue.

Interestingly, the experience so far has highlighted that the main problem is not the occasional lack of output, but regular *excess* output, more than the local, often rather weak or under-developed grid links could handle. In practice this has meant that output from wind plants has had to be 'curtailed' occasionally, i.e. not used, notably in the UK, US and China. This is wasteful, and undermines the economics of wind, although it is not that unusual in the energy generation field. For example, curtailment contracts are common for most generators on the UK grid, not just wind. They receive special compensatory payment if their output is not used. Clearly it would be preferable to strengthen the grids to avoid this problem, so that wind can displace high carbon sources on the grid more often, but that has costs. In the short term, curtailment may be cheaper.

Even with good grids, as wind capacity expands, there will come a point when there may occasionally be more deliverable output than is needed by users, i.e. when demand is low. To avoid curtailment, one reaction has been to reduce prices temporarily to stimulate use. This has happened recently in Germany and elsewhere with large wind capacity. Indeed, since wind plants run at almost zero marginal cost, and can produce electricity at very competitive prices compared to conventional fuel-dependent plants, at some times, when demand is low and wind high, some gas plants have become uneconomical to run. Basically wind can cut spot prices; see box 2.3.

Box 2.3. Wind cuts energy prices

A study by consultants Poyry, for the European Wind Energy Association, claimed that, primarily since it did not use fuel, wind power generation can reduce electricity prices by between €3 and €23/MWh, depending on the scale and location. It concluded that an increased penetration of wind power reduces wholesale spot prices: 'due to market dynamics and the lower marginal costs of wind power compared to conventional power, the spot electricity prices decrease by €1.9/MWh per additional 1000 MW wind capacity in the system' (Poyry 2010).

That already seems to have been playing out in practice. Poyry reported studies which suggest that, for example, in Germany, increased wind power has reduced costs in the range of €1.3–5bn per year. One result is that some power companies have begun to reduce charges to consumers. Another result is that some conventional plants have been running at a loss.

Offshore Wind Magazine (25 April 2010) noted that in Scandinavia, electricity retailers had taken steps to avoid this by encouraging generators to limit production and by implementing a minimum floor price. Similar arrangements have evidently been made in Australia, where excess wind generation has also at times been forcing pool prices to go negative, leading the market regulator to set a minimum floor price. But that is a pretty crude approach, basically fixing the market. It is not a sustainable long-term solution. In the same way, just letting prices fall may help avoid having to curtail the wind turbine output but, like curtailment itself, this does eat into profits, thereby limiting the potential for further investment. And since the spot price volatility is unpredictable, it makes it hard to plan ahead. So arguably there have to be limits on how low prices can fall, or else it becomes uneconomical to produce electricity, at least within a competitive market system.

I will be looking at this so-called 'merit order effect' on prices more closely in chapter 6. While these price reductions may be good for consumers, the loss of earnings may not be good for the energy system as a whole since, in the short term, wind will need gas plants for occasional back-up and, longer term, in addition to better grid links, extra grid balancing mechanisms will be needed to deal with lulls and excess output, all requiring investment. As I will be explaining later, the new balancing options include short- and long-term energy storage facilities, to make excess output ready to meet later demand peaks; smart grids, to allow for some load to be switched off/delayed when supply is low; and new inter-connectors to other grids in other countries, to allow for more exports and imports.

Balancing mechanisms may add slightly to the cost of wind and other variable renewables, but overall, as the review above should have indicated, wind power, offshore especially, seems to have good prospects, offering a major economically competitive resource with low environmental impacts, if carefully sited. Longer term, if on land and (much less likely) offshore sites get congested, there is even the prospect of airborne wind devices, perhaps supported by giant tethered kites or aerofoil sections and possibly operating in the high-altitude jet streams: a vast resource. The US government, amongst others, have been looking at this idea seriously (Energy Kites 2013, PES 2012).

The potential for novel, perhaps wild, ideas does not stop there. There have been designs for axially-rotating dirigible balloon-type devices, which could be used on land or offshore, and for huge on land vertical-axis rotors mounted on magnetic bearings, or vast cooling-tower-type cylinders, creating internal high-speed wind vortices, i.e. contained artificial tornados. Less aggressive, indeed quite charming, is the Windstalk design concept, an artificial forest of 50 m tall flexible stalks swaying in the wind, with the motion being converted into electricity partly by the compression of embedded piezo-electric material (Windstalk 2010).

Very large-scale and exotic devices seem unlikely to be viable technically, economically or environmentally, and I think that fairly standard designs will win out, although new ideas may yet offer new possibilities. It is an exciting area of innovation.

2.3 Wave power

Wind energy in various forms has been used for many centuries around the world, and in its modern version is spreading rapidly, but wave energy is a relative newcomer. When winds move over water they create waves through frictional effects. The waves persist for some time after the winds have died down. So wave power is really indirect wind energy, and can also *store* wind energy for a time. And since wind is indirect solar energy, waves are also solar driven. However, the fluid flow patterns associated with waves are very different from those associated with wind, so that the energy capture technology is very different. Rather than propeller-type turbines operating in basically laminar flow (albeit, with wind, of often changing direction), wave devices have to extract energy from the much more complex and ever-changing circular patterns of water movement that make up waves, whilst also coping with the risk of major storms.

The simplest device concept involves a floating buoy tethered to the seabed with the bobbing motion being used to drive some form of generator. Other devices, like the UK's

Oyster, have large hinged flaps mounted on the seabed, which sway back and forth under the influence of the wave motion. This flapping movement is used to pump a hydraulic fluid through a turbine to generate electricity, and in some near-shore designs, the turbine can be sited on land, which makes access for its maintenance easier (Oyster 2013).

Another concept is to have the waves moving up and down inside a part-filled vertical cylinder, open to the sea at the bottom, but constrained at the top, so that air trapped above the water is compressed by each upward water movement and pumped out through a turbine. On the down stroke, when the water falls, air is sucked back down through the turbine, so it has to be designed to be able to run in the reverse direction. Two-way Wells turbines are used in these devices, which are known as oscillating water columns (OWCs). OWCs are usually mounted on land at the edge of the sea, but some have been installed on floating pontoons, or further out to sea, raised from the seabed. One problem with OWCs is that the two-way Wells turbines are a compromised design and are not very efficient. However, they work reliably and this approach, developed by Wavegen in the UK, was one of the first to deliver grid power, with a 500 kW Limpit shoreline device installed in 2000 on the isle of Islay in Scotland (Wavegen 2013). There are many other possible configurations. Waves can be allowed to run up a floating ramp to overtop into a reservoir on a pontoon to create a head of water, which is then let out through a turbine. So it acts as a mini hydro plant, with a very low head (Wave Dragon 2011). Clearly this is a somewhat crude and not very efficient approach, but it is simple, and does allow operation far offshore.

There are clear benefits to going further offshore, where most of the wave energy is, although then the devices have to be designed to withstand major stresses. Some of the earlier designs attempted to absorb all or most of the energy in the impinging wave front. This can be problematic. After all, we have designed ships to avoid that; they usually try to cut through waves, rather than taking them broadside on.

So far, the most successful design for offshore use is the hinged, segmented, floating 'sea-snake'-like device, the UK's Pelamis system, which is anchored to the seabed from its nose, with the waves running down its flanks. The linked articulated segments pitch and heave under the impact of the waves and the resultant movements of the segments relative to each other drive hydraulic pumps feeding an onboard turbine generator. This concept means the device only uses a small fraction of the energy in the wave front, but that makes it more likely to survive strong waves and storms (Pelamis 2013).

The wave energy field is still in a state of creative flux, with many rival ideas under test around the world. Box 2.4 offers just a sample (Falcao 2010, Elliott 2012c).

In theory wave energy devices could collect energy from waves far offshore, but the cost of marine power cabling is high (around £1m/mile), so on land, inshore or near-shore location is more viable. The UK has some of the best sites for wave projects, off the north west of Scotland and Ireland, as well as off Cornwall, due to the strong westerly prevailing winds over the Atlantic. But there is also a good resource off Spain and Portugal.

In addition, studies have indicated that, if all the potential US wave and tidal resources could be used, they could in theory supply 30% of its electricity. Alaska and Hawaii are especially favoured for wave energy (DoE 2012, EPRI 2011). South America has a good wave resource. There is a 3 MW causeway-mounted wave plant in Mexico and plans for

Box 2.4. Wave power technology

Wave energy has come a long way since the early days in the 1970s and the work of pioneers like Stephen Salter and his nodding Duck (Ross 1995). There are now some more or less fully developed clear winners, like the UK's Pelamis wave snake (Pelamis 2013), the near-shore Oyster hinged wave flap (Oyster 2013) and Wavegen's shoreline-mounted OWC. There are also many buoy-type systems, like the US company OPT's Power Buoy (OPT 2013), and the Australian *CETO* system, which has an array of submerged buoys tethered to seabed pump units (CETO 2013). Ireland's Wavebob buoy system can be tuned to match different wave frequencies (Wavebob 2013), while some devices use waves to power hydraulic pistons directly, as with the Sea Dog pump developed in the US (Sea Dog 2013). In the UK Ecotricity is backing a bicycle pump-like Searaser piston device, which pumps seawater to a reservoir on a hill, so that electricity can be generated via a turbine when required (Ecotricity 2013).

Australian company Oceanlinx has a variant of the OWC with a variable pitch Wells turbine (Oceanlinx 2013). Finland's Wello Oy has a 500 kW Penguin device being tested at the European Marine Energy Centre (EMEC) in the Orkneys. It is a floating asymmetric vessel which houses an eccentric rotating mass, mounted on a vertical shaft (Wello 2013). And in the USA, Sea Ray uses motion constraint provided by internal magnets to extract energy from the pitch and heave of a floating device (Sea Ray 2012).

Ideas for smoothing out the energy absorbed from waves are also being explored. The floating Danish Waveplane has a series of slots in a ramp designed to catch waves at different heights, the captured flows then being used to create a vortex to drive a turbine (Waveplane 2004). In the UK, Ecotricity are backing the Snapper linear motor wave unit invented by Professor Ed Spooner at Edinburgh University. It has magnetically tripped springs, storing bursts of energy (Snapper 2010).The Archimedes wave swing compressed air piston device, invented in the Netherlands, and then taken on AWS Ocean Energy in the UK, has been developed into a multi-cellular ring version with hydraulic links between the cells to even out the energy flow (AWS 2013).

Many other devices are being developed, such as, in the USA, Atmocean's Wave Energy Sequestration Technology (WEST) buoy system (Atmocean 2013); in Portugal, the WEGA, pendulum-type 'gravitational wave energy absorber', by 'Sea For Life' (WEGA 2012); and in Australia, AquaGen's SurgeDrive using floating plates tethered to the seabed (SurgeDrive 2012). Scandinavian company Fred Olsen had earlier tested its Buldra system with floats supported underneath an oil-rig-type platform, but they are now focused on a smaller, ring-shaped Lifesaver device (REM 2012). Perhaps the most exotic design is the UK's Anaconda wave sock, a long flexible tube which makes use of differential pressures created inside by the bulge formed as water passes through the tube and then to a turbine (Bulge wave 2012).

projects in Colombia, Costa Rica, Guatemala, Panama and the Dominican Republic. As I have indicated, Australia is amongst the innovators in the field, and China and Japan have been looking seriously at wave energy. Indeed one of the first OWC projects was on a floating barge in Japan, the Mighty Whale (Ross 1995).

Although in terms of technology, the UK has made much of the running, and has installed some shoreline plants and near-shore prototypes, it has arguably been Portugal that has done best so far in terms of actually deploying devices on a significant scale. It installed three 750 kW Pelamis units in 2009, and has also been testing out other concepts, including Finland's 'wave roller' seabed-mounted hinged-flap system (AW Energy 2012).

The wave energy development path has not been without its problems, highlighting both the risks of operating in the sometimes very tough marine environment and the inevitably gradual process of learning how to deal with this via novel designs.

An early casualty, in 1995, was the 2 MW ART Osprey OWC, which was wrecked by a storm off the Scottish coast while being installed, before sand could be loaded into it as ballast. Finavera's prototype buoy system sank off the Oregon coast in 2007, just before its six-week test period ended. Finavera seems subsequently to have decided to focus instead on wind energy. In 2009 Trident's 80 tonne 20 kW oscillating linear motor prototype sank off East Anglia in the UK. In Australia, in 2010, a 2.5 MW Oceanlinx OWC prototype fell foul of heavy weather off the coast from Port Kembla, and was wrecked.

Much has been learnt from these episodes. They are part of the innovation and learning process, and the device teams persisted, developing more robust designs. Trident are installing a new device in Scotland (Trident 2012) and Oceanlinx are pressing ahead with a 1 MW Wave Energy Converter at Port Macdonnell, South Australia (Oceanlinx 2013). Although costs are still high compared to wind turbines, wave energy is following a similar process of innovative improvement. As technology improves and costs fall, with many new devices also under development, wave energy looks like being a significant addition to the list of viable renewable energy options.

2.4 Tidal power

As noted above, tidal mills were one of the first devices in which energy to drive machines was extracted from a renewable energy source, in this case the rise and fall of the tides in tidal rivers or estuaries. Small reservoirs were established which could be filled at high tide and then sealed off from the river, creating a head of water, much as with a low head or run of the river hydro plant. They were quite common in the Middle Ages in Europe and also in Asia and, like windmills, were used for grinding corn.

Much more recently, designs have emerged for electricity generation. The first large scheme, a 240 MW tidal barrage, was built on the Rance estuary in Brittany, Western France in the 1960s. Unlike traditional tidal mills, tidal barrages like this close off the entire estuary with a dam-like structure, although they have much lower heads than most hydro projects. The largest tidal rise available is in the Severn estuary in the south west of the UK which, at some locations and times, has a 10 m tidal range. Canada also has good sites, as does South Korea; indeed it now has a 240 MW barrage installed, and more are planned.

The tides are due to the gravitational pull of the Moon on the seas, modified by the pull of the Sun, sometimes acting in line, sometimes at right angles, so, as well as the daily tidal cycle due to the Earth's rotation, which leads to local ebbs and flows twice every 24 hours or so (but shifting slightly in time each day), there are high (spring) and low (neap) tides.

As with hydro plants, the energy available is proportional to the head height squared, so the bigger the better. The largest project so far seriously envisaged is an 8.6 GW scheme on the Severn estuary in the UK, between Cardiff and Weston-super-Mare, although there have been proposals for larger barrages further down the Severn estuary. Alternatively, smaller barrages on the Severn and elsewhere (e.g. the Mersey, Solway Firth) might be less invasive, and fully offshore artificial tidal lagoons even less so, since they would not block off the entire estuary/river. They would be totally self-enclosed, although lagoons that also enclose part of the river/estuary bank have been proposed (Swansea Bay 2013).

There are various operational options for these so-called 'tidal range' systems, i.e. tidal barrages and lagoons. Basically they trap high tides, letting water out through hydro-plant-type turbines to generate electricity. However, to increase output continuity, segmented lagoons, or barrages with lagoon attachments, could store water in separate basins, emptying it out in phased stages. Water can also be pumped 'uphill' behind barrages or into lagoons, using excess off-peak grid power. Operation on the incoming tidal flow as well as during the ebb is also an option, although that means using two-way turbines, which are more complex, expensive and prone to extra wear and breakdown.

Even so, with just ebb generation, tidal range systems might supply 15% of UK electricity (Yates *et al* 2013). Similar contributions might be made elsewhere. About 60 GW's worth of potential medium-to-large tidal barrage sites have been identified globally, as well as up to 100 GW of very large sites in Siberia (WEC 2001).

Although pumped storage and two-way operation are sometimes included as an option, most new barrage designs use conventional one-way turbines. The problem then is that they will only fire off twice roughly every 24 hours, with large pulses of energy for a few hours in each cycle. The peak tide timings also shift progressively each day. So the barrage output may not often match daily electricity demand peaks. The result is that, for example, although the 8.6 GW Severn Barrage might be able to generate 4.6% of UK's electricity, only some of that could actually be used effectively in practice, unless large amounts of money were also spent on major electricity storage facilities.

According to the generally pro-barrage Sustainable Development Commission, if built, the Severn Barrage would only reduce UK emissions by about 0.92%, not very much for the £20bn it was then expected to cost (SDC 2007), with the cost of electricity being seen as higher that almost any other energy supply option (Frontier Economics 2008). A 2010 UK Government review said there was little chance of support forthcoming, given the high cost (by then put at over £30bn) and the potentially large environmental impacts (DECC 2010). But it did say it would look at proposals from private developers.

In 2012, Hafren proposed a privately funded two-way ebb and flow barrage design for the Severn, operating on a lower head with a larger number of lower speed turbines. It claimed that the environmental impact would be lower and the output more regular, delivering power 16 h per 24 h cycle (Hafren 2013). Although the proposal has some political support (Hain 2013), some see it as a long shot, given the difficulties in raising large amounts of private capital and the strong and continuing environmental opposition to large barrages (Green Party 2012). In general, while in some locations, as South Korea has shown, smaller barrages might be viable (as may lagoons), large barrages look unlikely to be so, on economic grounds if nothing else (Elliott 2013).

2.4.1 Tidal current turbines

Smaller independent free-standing tidal turbines (essentially underwater wind turbines) have much less environmental impact than barrages and are faster and therefore cheaper to build. Large barrages could take ten years to complete during which time they would not be earning anything to pay back the capital borrowed to fund construction. A network of tidal current turbines around the coast could also deliver more nearly continuous output, since peak tide occurs progressively later in time at each site. They can also be designed to swivel around to run on the flow and the ebb, i.e. four times every 24 hours.

There are many tidal current projects and devices at various scales and stages of development. As with wave energy, there is a flurry of innovation. Indeed, the level of activity is possibly higher, since it is easier to extract energy from undersea tidal ebbs and flows than from waves. Wave devices tap into complex chaotic wave motions in the turbulent interface between air and water, whereas tidal devices avoid the turmoil on the surface, using the much smoother and more linear tidal energy flows. They are also much simpler than barrages. Tidal turbines run on the horizontal ebbs and flows, rather than on the vertical tidal range, so they do not need a dam to create a head of water, or a site where a dam could be built, just a location where there is fast-moving water, at least $2\ \mathrm{m\ s^{-1}}$ and more usually $4\ \mathrm{m\ s^{-1}}$. That usually means focusing areas where currents are speeded up due to topographical constrictions, as in channels or around islands.

The UK has some of the best sites for tidal current projects in the world. Scotland has 25% of Europe's tidal potential. Pentland Firth and Orkney waters contain six of the top ten UK sites and studies have suggested that the potential UK resource may actually have been understated by perhaps a factor of 10 (Salter 2008, MacKay 2007). It is now thought to be over 100 TWh, nearly 30% of UK's current electricity requirements (PIRC 2010).

Although, as with wave energy, the UK still leads in this field technologically, challenges are emerging from around the world. That is not surprising since, in addition to the UK, there are many locations worldwide where this technology can be used (Hardisty 2009). Other countries with large potentials include Canada (maybe 40 GW) and South Korea (perhaps up to 100 GW including barrages and lagoons). South Korea is planning several projects, some using turbines developed by Voith. Canada is also testing several designs (Fundy Force 2013). The US has a significant potential resource, with Maine being a favoured site (DoE 2011, 2012). Box 2.5 gives some examples of the many projects underway around the world.

As can be seen, tidal energy is a rapidly expanding field, and interest in tidal current power has clearly gone global, with a race to be first in what could be a very large global market. In the end what will probably decide the winners is economics. There is talk of some tidal current devices soon becoming competitive with offshore wind. Indeed, the UK Carbon Trust has said that, with continued targeted innovation, by 2025, 'the UK's best marine energy sites could generate electricity at costs comparable with nuclear and onshore wind' (Carbon Trust 2011).

As with wave and offshore wind devices, a key issue affecting cost is getting access to the turbines and generators for maintenance. In the extreme, units could be removed from site and brought back to shore, but most systems have arrangements for lifting the

Box 2.5. Tidal current technology and projects

The leader is the UK's 1.2 MW SeaGen turbine installed in Strangford Loch, Northern Ireland in 2010 (MCT 2013). It is feeding electricity to the grid under a Renewables Obligation contract. It has two propeller-like turbines mounted on a tower fixed to the seabed, although rather than swivelling round when the tidal direction changes, the turbines have reversible pitch capacity. Arrays of similar units, in 10 MW tidal farms, are planned off the Welsh and Scottish coasts.

There are many other horizontal axis propeller-type designs under development in the UK, including Swanturbines' Cygnet seabed-mounted device developed at Swansea University (Swanturbines 2013), Tidal Energy's DeltaStream device being tested off Wales (TEL 2013), ScotRenewables' floating pontoon with turbines mounted underneath (ScotRenewables 2013), Tidal Stream's extractable multiple turbine frame system (Tidal Stream 2013), and the device developed by Tidal Generation Ltd/Rolls Royce (TGL 2013). A novel variant is a two-bladed contra-rotating system developed initially at Strathclyde University (Nautricity 2013).

UK sites (such as EMEC in the Orkneys) are being used to test many devices including several from abroad, such as Hammerfest Strom's multi-bladed design (Hammerfest Strom 2012) and the Atlantis 1 MW double contra-rotating rotor. The latter deals with the change in direction of the tides by having two rotors designed for operation in opposite directions. One is feathered, while the other runs (Atlantis 2013). Although propeller-type designs have dominated so far, Irish developer Open Hydro have a novel 'open-centred' rim turbine device, with a series of 1 MW versions planned in France, Canada and Alderney in the Channel Islands (Open Hydro 2013). The UK's Pulse Tidal has developed a unique oscillating hydrofoil system, which it claims is suited to shallow water (Pulse Tidal 2013).

Vertical axis devices have also been developed, like the Kobold turbine tested by Ponte di Archimedes in the Strait of Messina, Italy (Calcagno and Moroso 2007) and Neptune's 'Proteus' ducted rotor, which was tested in the UK's Humber estuary (Neptune 2013). Although so far this concept seems to be less favoured, it still has some strong advocates, some of whom look to large projects. Canadian developer Blue Energy has been a pioneer in this field, with a vertical axis design mounted in a module. One early proposal was for large numbers to be joined together in a 2.2 GW 'tidal fence' causeway between islands in the Philippines (Blue Energy 2013). Some more recent UK designs also have a series of tidal turbines mounted in a cross-estuary 'tidal fence' or artificial 'tidal reef', in effect a *permeable* zero or very low head barrage (Tidal Fence 2008, REEF 2013).

There are many other innovative projects around the world. Voith Siemens have developed a gearless 1 MW propeller design, to be used in South Korea (Voith 2012). The Netherlands have the Tocardo rotor being tested at EMEC (Tocardo 2013), and many ideas are under test in North America, such as Canada's Clean Current ducted rotor (Clean Current 2012), Ocean Renewable Power Company's horizontal axis helical TidGen (ORPC 2013) and Verdant Power's propeller units, on test in New York's Hudson river (Verdant Power 2012). The US Department of Energy has allocated Marine Energy Grants to 14 projects, with more planned.

New Zealand also has plans for tidal current projects, possibly including Crest Energy's proposed $600m Kaipara Harbour project at Northland, with up to 200 turbines, each of 1.2 MW capacity. Australia also has some ambitious plans and proposals, including a 48 MW tidal project in Western Australia, and for large projects near Darwin in the Northern Territory, Port Phillip Heads in Victoria, and Banks Strait in Tasmania (Tidal Today 2011/12).

turbines' units out of the water for access at sea. For example, with the existing SeaGen, the blades can be raised hydraulically up the central tower support, while the new version of the SeaGen has a series of turbines mounted on a frame, attached to a giant arm, which lifts them to the surface. The Tidal Stream 'Triton' project has a similar arrangement, with a series of turbines mounted in a frame linked to arms swivelling up from a large gravity base on the seabed (Tidal Stream 2013).

Environmental impacts must of course be considered. All offshore systems have the potential for significant effects on marine wildlife, including dolphins, porpoises, grey seals and wildfowl, but it is argued that, compared with large tidal barrages, with high-speed turbines, relatively slowly rotating free-standing tidal rotors represent a low hazard. A range of environmental studies have been completed or are in hand, including assessment of the SeaGen's impact. So far no major problems seem to have emerged that cannot be limited by sensible design, location or mitigation measures. SeaGen has used a sonar system to monitor and warn of the approach of sea mammals (Phys Org 2012).

As with wave energy devices, operating at sea presents hazards and although, as I have said, in general extracting energy from tides seems easier than working with waves, some tidal prototypes have failed, for example, Open Hydro's 1 MW test project in Nova Scotia. This was installed in 2008, but had to be extracted a year ahead of schedule. Evidently it lost all its blades in rough water. However, Open Hydro have learnt from this and are pushing on with an 8 MW array off the French coast, backed by EDF (Open Hydro 2013). Atlantis had some initial problems with poorly fabricated blades, but these are just the sort of teething troubles that can be expected with new technology. As can be seen, there is a flurry of activity underway in this field, making it hard to see which designs and which countries will win out (Elliott 2012d). Although South Korea has ambitious plans and the USA/Canada may follow, for the moment the EU leads. On the basis of the European Renewable Energy Action Plan, the EU should have 2.1 GW in place by 2020, 1.3 GW in UK waters, 380 MW off France, 250 MW off Portugal, 100 MW off Ireland, 75 MW off Spain and 3 MW off Italy. Whether all of this will be achieved on time remains to be seen. For example, only around 1 GW of sites have been agreed so far in the Pentland Firth area and only some may be developed by 2020.

Primarily lunar-driven tidal currents are of course not the only form of energy flow in the seas. There are also very large non-tidal movements like the Gulf Stream, due ultimately to solar heating of the seas, which creates massive flows of water around the planet. Some of the huge amount of energy in these ocean streams could be tapped in some locations and there are some projects looking at that, notably off Florida, where the resource is said to be up to 10 GW (Ansari 2009).

There is sometimes some terminology confusion, in that the terms 'tidal current' (the term I use) and 'tidal stream' are often used interchangeably (which is fair enough), but they are also both sometimes used to cover non-tidal ocean currents/streams. Worse still is the commonly used term 'tidal waves', which are neither tides or waves. They are rapid flows in seas due to seabed earthquakes, nowadays, tragically after some awful disasters, more usually called tsunamis. Needless to say, they have no potential as energy sources. But then there is plenty of energy in tidal currents and ocean streams and tidal lagoons could well prove viable in some locations.

2.4.2 The design options

Tidal barrages, large or small, are likely to have similar basic designs, but some of their environmental impacts will vary depending on the turbines used, the location and operational patterns. In terms of lagoons, the basic physics is the same as for barrages, but there are two basic options, fully offshore and partially offshore. Fully offshore lagoons in wide, open channels will have low impacts, since they will not impede tidal ebbs and flows, but although that would also be true of lagoons that include estuary banks and, although these will cost less to build, this configuration can lead to impacts on the more ecologically sensitive shoreline areas. Either type can be operated on ebbs as well as flows, and can also be segmented, to allow for judicious balancing of levels and outputs from separate basins, possibly with the addition of pumping using off-peak electricity, to get more continuous grid supply outputs.

In terms of tidal current turbines, conventional propeller-type designs, such as SeaGen, seem likely to dominate, and supporters of this approach are sometimes a little dismissive of rival design concepts. For example, oscillating hydroplane designs have a horizontal aerofoil section mounted on an arm, pushed up and down by the tidal flow. It is sometimes joked that the Victorians found out that (propeller) screws were better than paddle wheels. Although hydrofoil systems are not really paddles (and tidal turbines are not propellers), hydrofoils do have the problem that they do not absorb energy at the end of each up and down stroke (when they stop and reverse direction), only during the middle of each cycle. But they can have large cross sections and, unlike propellers, when they are moving the aerofoil section is used fully for energy extraction. With propellers, most of the energy take is from the faster moving outer tip section. The basic physics of tidal current turbines is similar to that of wind turbines, but with topographical constraints on flow rates (Hardisty 2009). Since energy out is proportional to the water speed cubed there is a premium on high-speed water flow. Some designs use ducts or cowls surrounding the rotor to accelerate the water flow (Lunar Energy 2013). The same idea has been tried with wind turbines, with some success. A 'Wind Lens' wind concentrator is being developed in Japan, with a two- or three-fold energy output augmentation claimed (Ohya and Karasudani 2010). A problem with ducting for wind turbines is that the whole assembly has to turn as wind directions change. If the rotors can be reversed, that is not a problem with tidal devices, since tidal currents stay in the same line, ebbing and flowing back and forward. However the advantage of ducts, in terms of extra water speed, may be offset by the extra cost of the duct material and this concept now seems to be less favoured. For example, although interest in the idea continues, Neptune's Proteus tidal turbine, which had extensive ducting, has had problems and this design has been abandoned (Steiner-Dicks 2013). It also had vertical axis blades but, it is claimed, unlike with variable direction wind, vertical axis designs do not have the same advantage for use with tidal currents, given that they ebb and flow along the same line.

Negative views are also sometimes heard concerning small devices aiming to capture energy at sites with low flow rates. The square law on blade size and cube law on flow rate would seem to make this uneconomic although, to be fair, the low-speed resource is much larger than the high-speed resource, and cheap, simple, robust low-speed devices might find a role, since they could be used in a much wider range of sites (Hales 2013).

In that context, one of the most intriguing ideas is the tidal kite being developed by Swedish company Minesto. An aerofoil wing, with a rotor and generator mounted on it, is tethered to the seabed but free to move in the tidal flow, under rudder control, at enhanced speed in a figure-of-eight pattern. It is being tested off Northern Ireland. Cable stresses are an obvious issue, since the cable also has to carry the electrical connection. However, the developers claim that, since the flow over the rotor blades is speeded up by perhaps 10 times, the system could extract power economically from relatively low tidal flow rates, thus, in effect, expanding the potential tidal resource (Minesto 2013).

Another novel design makes use of the Venturi effect, tapping off tidal flows with the primary water flow being used to create a faster secondary flow in a smaller tributary pipe. It is claimed that this approach can be used at a variety of scales, including in large tidal fences and low flow rate sites, and will reduce environmental impact (VerdErg 2009). Clearly new ideas continue to emerge, and although propeller-type designs seem likely to win out, in the tidal current field, there is evidently still all to play for.

2.5 Power in perspective

All the mechanical power systems considered in this chapter use fluid flows, though in different ways. Water is around 800 times the density of air, so the rotors used to capture energy from water flows can be smaller than those used for air flows, for the same energy capture. However, against that, tidal current water speeds are usually much less than the wind speeds used for wind turbines. With hydro and tidal range projects, high-speed water flows are created from the potential energy of water trapped by a barrage or lagoon wall, so the turbines are usually relatively small, but have high power ratings.

These various flow rates and rotor configurations will lead to differing energy conversion efficiencies, although efficiency is a difficult concept to apply to the use of renewable sources like this, since the incoming energy is essentially free and unlimited. That contrasts with the fuels for conventional power plants, where efficiency can be measured as the well-defined energy content of the fuel divided by the energy value of electricity produced.

Conventional fossil- and nuclear-fuelled power plants, using heat to raise steam to drive turbines, have basic energy conversion efficiencies of around 30–35%, due to the thermodynamic cycle limits associated with such systems. That said, it is possible to recycle some of the waste heat rejected from the turbines, as in combined cycle gas turbine plants (CCGTs), which can raise the overall efficiency to 50% or more, by using the hot exhaust from the first turbine to raise steam for a second turbine. Even better, operation in combined heat and power (CHP) mode, using the waste heat as well as power, can at least double the overall energy conversion efficiency. I will look at CHP (sometimes also called cogeneration) in section 3.4. Efficiencies of 70–80% are possible.

Even so, these are still steam-raising thermal plants and the basic energy conversion process is still limited by thermodynamic laws. By contrast, mechanical devices like wind turbines have a different theoretical energy conversion efficiency limit, known as the Betz limit, of 59.3% of the available incident energy. It is based on the process of converting the energy in fluid movement to energy in rotor blade movement. In practice, wind or tidal turbines cannot achieve anything like 59.3% efficiency (although 45% energy extraction has been claimed for the SeaGen tidal turbine), but all that means in design terms is that larger rotors or better sites are needed to get the same output as from a device which could attain 59.3% (Vennell 2013).

Crucially, wind and wave energy, in common with solar energy (from which they are derived) are variable and weather dependent, while tidal energy is predictable but cyclic, linked (mainly) to the lunar cycle. Earlier I mentioned what these energy output variations mean for wind energy projects in terms of curtailment (when there was too much output), and I will be looking at how variability, including shortfalls, can be dealt with in chapter 5. For the moment, however, I will focus just on how source variability can be reflected numerically. Given that it reduces the total amount of energy available over time, it can be presented by assigning 'load factors' to the technology, indicating the percentage of its full rated power a plant can deliver over a year. They are also sometimes called 'capacity factors', and though there are slight differences, the basic measure is the percentage of energy actually produced of the theoretical maximum that could be produced if the device ran at full power continuously, usually averaged over a year.

As I have indicated, hydro plants achieve a global average load factor of around 44%, although some can do much better than that, e.g. in the range 70–80%. Wind projects are also very site and region specific, as well as being local-weather dependent. UK on land wind load factors vary from below 20% to above 40% depending on site; 30% is often taken as an average, but is moving up as the technology improves. Offshore wind is sometimes quoted as 35–40%, but one new Danish offshore wind farm has achieved 47%. That should improve with sites further out, new designs and more operating experience. Since there are fewer wave and tidal projects in operation, it is harder to come up with reliable load factor estimates, but some UK estimates are presented in table 2.2, from the 2050 Energy Pathways modelling exercise carried out by DECC. As can be seen, DECC puts the load factors for wave and tidal range projects at 23%, and for tidal current projects at 36%. It also puts the load factor for offshore wind at 45%. Its estimates for the load factors for some of the other heat-based renewables are much higher, with biomass plants offering more or less continuous output from 'firm' capacity, which I will be looking at in the next chapter.

Note that all energy generation projects have variable outputs, due to unscheduled down time, plant and grid failures, maintenance periods, occasional fuel shortages and so on. For example, the 2011 Digest of UK Energy Statistics put UK nuclear load factors as 69.3% (in 2006), 59.6% (2007), 49.4% (2008), 65.6% (2009) and 59.4% (2010). That averages out at 60%. The Nuclear Energy Institute quotes average US nuclear plant load factors (1971–2009) as 70%, although the World Nuclear Association puts the current average US load factor at 87%. Elsewhere they are often lower. For example some of the (now closed) Japanese plants only achieved around 44% (WNA 2012).

Table 2.2. Annual load factors for renewable energy technologies in the UK.

Offshore wind	45%
On land wind	30%
Wave	23%
Tidal stream	36%
PV solar	10%
Hydro	38%
Biomass—electricity from CHP	90%
Geothermal—electricity from CHP	80%
Fuels from biomass	90%
Solar hot water	50%
Bioenergy, energy from waste	80%
Tidal range	23%

DECC (2012a). CHP = operated for heat production as well as electricity.

While nuclear proponents claim that much higher figures will be attained in future (over 90%), load factors for renewables may also rise and, as noted above, for some of the technologies I have not yet looked at (biomass and geothermal CHP particularly) they can be very high: see table 2.2. As indicated above, they are much like conventional 'firm' energy sources.

Even so, clearly most of the renewables looked at in this chapter are significantly affected by the variability of their energy sources. Note I do not use the term 'intermittency'. All power plants operate intermittently to a degree, as indicated above. The grid system copes with this by providing backup and balancing to ensure continuity of supply. In that context, as chapter 5 will illustrate, the extra variability due to the use of weather and time-related renewable sources may not be such a large problem as it may first seem. There are ways to deal with it, although they may add to the cost, as will strengthening grids to make use of these often remotely located renewable sources.

At the same time, their geographical distribution may help deal with local variability, if the projects can be linked up across wide areas via supergrids. For example, the generally prevailing westerly winds in north west Europe mean that the UK and Ireland get the wind fronts first, Germany and central Europe some hours or even days later. Judicious electricity trading, as well as taking advantage of regional clock time differences, can help balance peaks and troughs, with wind forecasting playing a key role. That is now quite a well-developed science, and allows for the timely run-up of any back-up plant needed, or pre-planning of demand management activities, thus reducing balancing costs.

There are of course many other factors affecting costs. I will be looking at some of them, and at cost comparisons, in chapters 5 and 6. For the moment suffice it to say that large established hydro is one of the cheapest sources of electricity available, on land wind is at present the cheapest of the main new renewables, while costs for the offshore wind, and also the so-called marine options, wave and tidal, are falling.

Although some UK scenarios have marine renewables supplying more energy than wind by 2050 (UKERC 2013), at present wind is in the lead, in terms of the main new (non-hydro) renewables looked at in this chapter. That seems likely to remain the case around the world, given the much larger wind resource, especially if offshore floating wind systems prove to be viable. As I have indicated, environmental impacts for wind, wave and tidal current devices are generally low, but there can be significant issues with large hydro and tidal barrages. So again, it seems likely that wind, and possibly wave and tidal current systems, will dominate although, as later chapters will suggest, some of the other renewables may well challenge them longer term, PV solar especially.

Summary points

- **Hydroelectric power** is a major existing renewable energy source, but there are question marks about the environmental impacts of large hydro projects.
- **Windpower** is expanding rapidly on land and increasingly offshore, and is the cheapest of the major new renewables, with low environmental impacts if properly sited.
- **Wavepower** is developing rapidly around the world, despite it being hard to extract energy in sometimes very rough marine environments.
- **Tidal barrages** may be technically viable in some locations, but there are environmental and economic problems; lagoons and, more especially, **tidal current turbines** seem more likely to prosper. There are many tidal current projects underway.
- **Variability** is an issue but may not be as hard to deal with as is sometimes thought—there are many grid-balancing options, although they may add to the cost.

References

4cOffshore 2013 Global data base for offshore wind projects, http://www.4coffshore.com

Ansari A 2009 Is the ocean Florida's untapped energy source?, CNN.comTechnolgy, July 27, http://edition.cnn.com/2009/TECH/07/27/ocean.turbines/index.html

Arup 2010 ARUP news coverage of Aerogenerator X project, http://www.arup.com/Home/News/2010_07_July/27_Jul_2010_Arup_and_Wind_Power_Limited_unveil_10MW_Wind_Turbine.aspx

Atlantis 2013 Atlantis Resources company website, http://www.atlantisresourcescorporation.com

Atmocean 2013 Ocean HydroPower System website, http://www.atmocean.com/

AU *et al* 2009 Position statement on hydro by The African Union, The Union of Producers, Transporters and Distributors of Electric Power in Africa, The World Energy Council, The International Commission on Large Dams, The International Commission on Irrigation and Drainage, and The International Hydropower Association

AW Energy 2012 Wave Roller company website, http://aw-energy.com/

AWS 2013 AWS Ocean company website, http://www.awsocean.com/technology.aspx

Bach, P 2011 Norway Preparing for Balancing European Wind Power, The Oil Drum website, Feb 7, http://www.theoildrum.com/node/7404

Baerwald E, Edworthy J, Holder M and Barclay R 2008 A large-scale mitigation experiment to reduce bat fatalities at wind energy facilities *J. Wildl. Manag.* **73** (7)
Science Daily, http://www.sciencedaily.com/releases/2009/09/090928095347.htm

Bayar T 2013 Marine life unhindered by offshore wind farm, study says *Renewable Energy World*, May, http://www.renewableenergyworld.com/rea/news/article/2013/05/offshore-wind-farm-has-lack-of-significant-impact-on-marine-life-study-finds?cmpid=WNL-Wednesday-May22-2013

Blue Energy 2013 Blue Energy company website, http://www.bluenergy.com
Blue H 2013 Blue H offshore wind turbine company website, http://www.bluehgroup.com/
BRE 2007 Micro-wind turbines in the urban environment, Building Research Establishment, Garston. See also A review of micro-generation and renewable energy technologies, NHBC/BRE, http://nhbcfoundation.org/Projects/tabid/54/Default.aspx
Bulge wave 2012 Bulge Wave Anaconda website, http://www.bulgewave.com/
Calcagno G and Moroso A 2007 The Kobold marine turbine: from the testing model to the full scale prototype, Tidal Energy Summit, London, November 28–29, http://www.tidaltoday.com/tidal07/presentations/GuidoCalcagnoMoroso.pdf
Carbon Trust 2011 Accelerating marine energy, Carbon Trust report CTC797, Carbon Trust, London, July, http://www.carbontrust.co.uk/news/news/press-centre/2011/Pages/MEA.aspx
CEFAS 2004 Offshore wind farms Guidance Note for Environmental Impact Assessment in respect of FEPA and CPA requirements, June 2004, Centre for Environment, Fisheries and Aquaculture Science for the Marine Consents and Environment Unit, DEFRA/DfT, London, http://www.cefas.co.uk/publications/files/windfarm-guidance.pdf
CETO 2013 Carnegie Wave Energy company website, http://www.carnegiewave.com/
Clarke A 2012 Environmental Impacts of Renewable Energy, PhD Thesis, Open University, Milton Keynes
Clean Current 2012 Clean Current company website, http://www.cleancurrent.com/
Cummings J 2012 Wind farms and health: it's not black or white *Renewable Energy World*, Feb 17, http://www.renewableenergyworld.com/rea/news/article/2012/02/wind-farms-and-health-its-not-black-or-white?page=2
DECC 2010 Severn Tidal Power: Feasibility study conclusions, Department of Energy and Climate Change, London, http://www.decc.gov.uk/assets/decc/What%20we%20do/UK%20energy%20supply/Energy%20mix/Renewable%20energy/severn-tp/621-severn-tidal-power-feasibility-study-conclusions-a.pdf
DECC 2012a 2050 Pathways Calculator, Department of Energy and Climate Change, London, http://webarchive.nationalarchives.gov.uk/20121217150421/, http://decc.gov.uk/en/content/cms/tackling/2050/2050.aspx
DECC 2012b DECC Public Attitudes Tracker-Wave 2, UK Department of Energy and Climate Change, London, http://www.decc.gov.uk/assets/decc/11/stats/6410-decc-public-att-track-surv-wave2-summary.pdf
Deepwater Wind 2012 US offshore wind proposals, http://www.dwwind.com
DoE 2011 Assessment of Energy Production Potential from Tidal Streams in the United States, Georgia Tech Research Corporation report for the US Department of Energy, Washington DC, http://www1.eere.energy.gov/water/news_detail.html?news_id=18017
DoE 2012 DOE Reports Show Major Potential for Wave and Tidal Energy Production Near U.S. Coasts, US Department of Energy news release, Jan 18, http://www1.eere.energy.gov/water/news_detail.html?news_id=18017
Donald R 2012 Wind turbine syndrome: who's doing the research? *Carbon Brief*, Oct 24, http://www.carbonbrief.org/blog/2012/10/wind-turbine-syndrome-whos-doing-the-research
Ecotricity 2013 SeaRaser wave device, Ecotricity company website, https://www.ecotricity.co.uk/our-green-energy/our-green-electricity/and-the-sea/seamills
EIA 2013 US Department of Energy, Energy Information Administration, http://www.eia.gov/cfapps/ipdbproject/iedindex3.cfm?tid=6&pid=37&aid=12&cid=regions&syid=2007&eyid=2011&unit=BKWH

Elliott D 2009 Sounding off on wind turbines *Renew Your Energy* blog, Environmental Research Web, Aug 21, http://environmentalresearchweb.org/blog/2009/08/sounding-off-on-wind-turbines.html

Elliott 2010 Wind turbine noise impacts *Renew Your Energy* blog, Environmental Research Web, Feb 6, http://environmentalresearchweb.org/blog/2010/02/wind-turbine-noise-impacts.html

Elliott D 2012a Bashing wind *Renew Your Energy* blog, Environmental Research Web, July 28, http://environmentalresearchweb.org/blog/2012/07/bashing-wind.html

Elliott D 2012b New offshore wind technology *Renew Your Energy* blog, Environmental Research Web, Sept 8, http://environmentalresearchweb.org/blog/2012/09/new-offshore-wind-technology.html

Elliott D 2012c New wave energy technology *Renew Your Energy* blog, Environmental Research Web, Sept 15, http://environmentalresearchweb.org/blog/2012/09/new-wave-energy-technology.html

Elliott D 2012d New tidal turbine ideas *Renew Your Energy* blog, Environmental Research Web, Sept 22, http://environmentalresearchweb.org/blog/2012/09/new-tidal-turbine-ideas.html

Elliott D 2013 A silly tidal idea? *Renew Your Energy* blog, Environmental Research Web, April 5, http://environmentalresearchweb.org/blog/2013/04/a-silly-tidal-idea.htmlApril

EPRI 2011 Mapping and Assessment of the US Ocean Wave Resource, Electric Power Research Institute, Palo Alto, http://www1.eere.energy.gov/water/pdfs/mappingandassessment.pdf

ESHA 2013 European Small Hydro Association, Brussels, http://www.esha.be

EST 2008 Power in Numbers Energy Saving Trust, London, http://www.energysavingtrust.org.uk/Publications2/Local-delivery/Legislation-and-policy/Power-in-numbers-summary-report

Energy Kites 2013 Airborne wind power website, http://energykitesystems.net

Falcao A 2010 Wave energy utilization: A review of the technologies *Renew. Sust. Energy Rev.* **14** 899–918

FREDS 2008 'Scottish Hydropower Resource Study' commissioned for the Forum for Renewable Energy Development in Scotland (FREDS) from Nick Forrest Associates Ltd, The Scottish Institute of Sustainable Technology (SISTech), and Black & Veatch Ltd, http://www.scotland.gov.uk/Topics/Business-Industry/Energy/Energy-sources/19185/Resources/17613/FREDSHydroResStudy

Frontier Economics 2008 Analysis of a Severn Barrage, Consultants report for WWF *et al*, http://assets.wwf.org.uk/downloads/frontier_economics_barrage_repo.pdf

Fundy Force 2013 Fundy Ocean Research Center for Energy: Canadian tidal stream energy information portal, http://fundyforce.ca/

Gatto K 2011 Vertiwind: Floating wind turbine project launched *Phys. Org.*, Feb 7, http://phys.org/news/2011-02-vertiwind-turbine.html

Green Party 2012 Severn Tidal Power Briefing, the UK Green Party, https://docs.google.com/file/d/0B5-HODLqTnX7OXAyR2x2XzBYZ0E/edit?pli=1

GWEC 2013 GWEC Global Wind Statistics 2012, Global Wind Energy Council, Brussels, http://gwec.net

Hafren 2013 Hafren Power Severn Barrage proposal, http://www.hafrenpower.com

Haggett C 2012 The social experience of noise from wind farms, ed J Szarka, R Cowell, G Ellis, P Stracham and C Warren *Learning from Wind Power* (Basingstoke: Palgrave Macmillan)

Hain P 2013 New Severn Barrage would exploit two way tides, Click On Wales website, Institute of Welsh Affairs, Jan 11, http://www.clickonwales.org/2013/01/new-severn-barrage-would-exploit-two-way-tides/

Hales 2013 Hales tidal turbine company website, http://www.hales-turbine.co.uk/

Halliday 2013 Micro Hydro screw company website, http://www.hallidayshydropower.com/
Hammerfest Strom 2012 Hammerfest Strom company website, http://www.hammerfeststrom.com
Hardisty J 2009 *The Analysis of Tidal Stream Power* (London: Wiley)
Harvey D 2010 *Carbon-Free Energy Supply* (London: Earthscan) pp 306–8
Hathaway T 2008 International Rivers Network, quoted in the *The Guardian* 21/4/08, http://www.irn.org
Hexicon 2013 Hexicon company website, http://www.hexicon.eu
Hyperwind 2013 EU Hyperwind project website, http://www.hyperwind.eu
Hywind 2012 Hywind floating wind turbine, Statoil website, http://www.statoil.com/en/Technology Innovation/NewEnergy/RenewablePowerProduction/Offshore/Hywind/Pages/HywindPutting WindPowerToTheTest.aspx
IHA 2002 The International Rivers Network statement on GHG emissions from reservoirs, a case of misleading science, report by Luc Gagnon, International Hydropower Association, Sutton, UK
IHA 2013 Sustainable Hydropower website, http://www.sustainablehydropower.org
IPSOS 2011 IPSOS Global Advisor, global poll carried out in May, published in June, http://www.ipsos-mori.com/Assets/Docs/Polls/ipsos-global-advisor-nuclear-power-june-2011.pdf
JRC 2013 Assessment of the European potential for PHS, European Commission Joint Research Centre, http://setis.ec.europa.eu/newsroom-items-folder/jrc-report-european-potential-pumped-hydropower-energy-storage
JWPA 2010 Long-Term Installation Goal on Wind Power Generation and Roadmap V2.1, Japan Wind Power Association, http://jwpa.jp/page_132_englishsite/jwpa/detail_e.html
Lenz 2012 Over 60% renewable electricity country list, Lenz Blog, Karl-Friedrich Lenz, http://k.lenz.name/LB/?p=6525
Lindeboom H, Kouwenhoven H, Bergman M, Bouma S, Brasseur S, Daan R, Fijn R, de Haan D, Dirksen S, van Hal R, Hille Ris Lambers R, ter Hofstede R, Krijgsveld K, Leopold M and Scheidat M 2011 Short-term ecological effects of an offshore wind farm in the Dutch coastal zone; a compilation *Environ. Res. Lett.* **6** 035101
Lunar Energy 2013 Lunar energy company website, http://www.lunarenergy.co.uk
MacKay D 2007 Under-estimation of the UK Tidal Resource, University of Cambridge, http://www.inference.phy.cam.ac.uk/sustainable/book/tex/TideEstimate.pdf
MCT 2013 Marine Current Turbines (now part of Siemens) company website, http://www.marineturbines.com/
Minesto 2013 Minesto's 'Deep Green' Tidal Kite company website, http://www.minesto.com
Mott MacDonald 2011 Costs of low-carbon technologies, report for the Committee on Climate Change, May, http://hmccc.s3.amazonaws.com/Renewables%20Review/MML%20final%20report%20for%20CCC%209%20may%202011.pdf
Nautricity 2013 Nautricity company website, http://www.nautricity.com
Neptune 2013 Neptune company website, http://www.neptunerenewableenergy.com
Oceanlinx 2013 Oceanlinx company website, http://www.oceanlinx.com/
Offshore Grid 2011 Offshore Grid: Offshore Electricity Infrastructure in Europe, 3E (coordinator), Dena, EWEA, ForWind, IEO, NTUA, Senergy, SINTEF, http://www.offshoregrid.eu
Ohya Y and Karasudani T 2010 A shrouded wind turbine generating high output power with wind-lens technology, *Energies* **3** (4) 634–49
Open Hydro 2013 Open Hydro company website, http://www.openhydro.com
OPT 2013 Ocean Power Technologies company website, http://www.oceanpowertechnologies.com
ORPC 2013 Ocean Renewable Power company website, http://www.orpc.co/

Oyster 2013 Aquamarine Power, Oyster wave power technology company website, http://www.aquamarinepower.com/technology

Pelamis 2013 Pelamis wave energy company website, http://www.pelamiswave.com/

PES 2012 'Power and Energy Solutions' review of Skymill Energy's airborne device, http://www.pes.eu.com/wind/the-skys-no-limit/2075/

Phys Org 2012 First seabed sonar to measure marine energy effect on environment and wildlife *Phys. Org.*, July 10, http://phys.org/news/2012-07-seabed-sonar-marine-energy-effect.html#jCp

PIRC 2010 Offshore valuation, Offshore Valuation Group, led by the Public Interest Research Centre, Machynlleth, http://www.offshorevaluation.org

Plumer B 2012 Are wind turbines making people sick? Or is it all psychological? *Washington Post*, Oct 23, http://www.washingtonpost.com/blogs/wonkblog/wp/2012/10/23/are-wind-turbines-making-people-sick-or-is-it-all-just-psychological/

Poyry 2010 Wind Energy and Electricity Prices, Poyry consultants report for the European Wind Energy Association, http://www.ewea.org/fileadmin/ewea_documents/documents/publications/reports/MeritOrder.pdf

PRNewswire 2013 China's offshore wind market expected to grow to US$16 billion, Your Industry News *PRNewswire*, Feb 28, http://www.yourindustrynews.com/china%27s+offshore+wind+market+expected+to+grow+to+us$16+billion_87489.html

Pulse Tidal 2013 Pulse Tidal company website, http://www.pulsegeneration.co.uk

Qi W 2011 China plans $1.8 trillion wind power plan for 2050 *Windpower Monthly*, Oct 20, http://www.windpowermonthly.com/article/1099715/China-plans-18-trillion-wind-power-plan-2050

Qi W 2013a China aiming to take grid-installed capacity to 75 GW *Windpower Monthly*, Feb 28, http://www.windpowermonthly.com/article/1172970/China-aiming-grid-installed-capacity-75GW

Qi W 2013b Analysis—Chinese wind curtailments double in 2012 *Windpower Monthly*, Feb 28, http://www.windpowermonthly.com/article/1171987/Analysis—Chinese-wind-curtailments-double-2012

REEF 2013 Severn Tidal Reef Project, http://www.severntidal.com/

REM 2012 Fred Olsen installs first wave device at Falmouth's wave energy test site *Renewable Energy Magazine*, April 13, http://www.renewableenergymagazine.com/article/fred-olsen-installs-first-wave-device-at

RISO 2013 Deep wind floating offshore wind project, RISO, http://www.deepwind.eu

Ross D 1995 *Power from the Waves* (Oxford: Oxford University Press)

RT 2013 Japan to start building world's biggest offshore wind farm this summer *Russia Today*, Jan 19, http://rt.com/news/japan-renewable-energy-resource-290/

Salter S 2008 Renewable electricity generation technologies, Fifth Report of Session 2007–08, Volume II, HC216-II, page Ev 106, House of Commons, Innovation, Universities, Science and Skills Committee, London, The Stationary Office, http://www.publications.parliament.uk/pa/cm200708/cmselect/cmdius/216/216ii.pdf

Schmoeller H 2006 From Interconnected Hydro Power Plants to Trading Recommendations: Possible Applications for Stochastic Programming, Power Point slides from ÖGOR Energy Workshop (EW06) Pricing Models for Electricity Markets, Vienna

ScotRenewables 2013 ScotRenewables company website, http://www.scotrenewables.com

SDC 2007 Turning the Tide: Tidal Power in the UK, Sustainable Development Commission, London, http://www.sd-commission.org.uk/publications/downloads/Tidal_Power_in_the_UK_Oct07.pdf

SeaDog 2013 IRNS company website, http://inri.us/index.php/SEADOG

SeaRay 2012 Columbia Power Technologies company website, http://www.columbiapwr.com/technology.asp

Smith J, Nayak D and Smith P 2013 Renewable energy: avoid constructing wind farms on peat *Nature* **489** 33

Snapper 2010 Snapper wave device website, http://www.snapperfp7.eu/

Sourcewatch 2010 Comparative electrical generation costs, Sourcewatch website, data from California Energy Commission, http://www.sourcewatch.org/index.php?title=Comparative_electrical_generation_costs

Sovacool 2009 Contextualizing avian mortality: a preliminary appraisal of bird and bat fatalities from wind, fossil-fuel, and nuclear electricity *Energy Policy* **37** 2241–8

Steiner-Dicks K 2013 Neptune Renewable Energy: chosen approach technically flawed; plans for liquidation *Tidal Today*, Feb 13, http://social.tidaltoday.com/technology-engineering/fortnightly-intelligence-brief-31-january-%E2%80%93-13-february-2013?utm_source=Newsletter1501&utm_medium=Newsite%2B&utm_content=1501&utm_campaign=Tidal

SurgeDrive 2012 SurgeDrive wave device, AquaGen Technologies website, http://www.aquagen.com.au/technologies/surgedrive

Swansea Bay 2013 Swansea Bay Tidal Lagoon project, company website, http://www.tidallagoonswanseabay.com/

Swanturbines 2013 Swanturbines company website, http://www.swanturbines.co.uk/

Sway 2013 Norwegian Sway floating offshore wind turbine design, http://www.sway.no/

Taylor D 2012 Wind energy *Renewable Energy*, ed G Boyle (Oxford: OUP)

TEL 2013 Tidal Energy Ltd company website, http://www.tidalenergyltd.com/

TGL 2013 Tidal Generation Ltd company website, http://www.tidalgeneration.co.uk/

Tidal Fence 2008 Severn Tidal Fence, IT Power, http://www.itpower.co.uk/node/141

Tidal Stream 2013 Tidal Stream Triton device company website, http://www.tidalstream.co.uk/

Tidal Today 2011/12 Website, London, http://www.tidaltoday.com

Tocardo 2013 Tocardo tidal turbine company website, http://www.tocardo.com/

Trident 2012 Developers website: http://www.tridentenergy.co.uk

UKERC 2010 Great Expectations, UK Energy Research Centre, London, http://www.ukerc.ac.uk/support/tiki-index.php?page=Great+Expectations%3A+The+cost+of+offshore+wind+in+UK+waters

UKERC 2013 The UK energy system in 2050: Comparing Low-Carbon, Resilient Scenarios, UK Energy Research Centre, London, http://www.ukerc.ac.uk/support/tiki-download_file.php?fileId=2976

Vennell R 2013 Exceeding the Betz limit with tidal turbines *Renew Energy* **55** 277–85

Verdant Power 2012 Verndant Power company website, http://www.verdantpower.com

VerdErg 2009 SMEC Spectral marine energy converter, VerdErg, http://www.verderg.com/attachments/-01_SMEC_Doc_Oct%2009.pdf

Voith 2012 Voith Tidal device, company information, http://voith.com/en/Voith_Ocean_Current_Technologies(1).pdf

Wave Dragon 2011 Wave Dragon website, http://www.wavedragon.net

Wavebob 2013 Wavebob company website, http://www.wavebob.com

Wavegen 2013 Wavegen, now part of Voith Hydro, company website, http://www.wavegen.co.uk/

Waveplane 2004 WavePlane, WPP A/S website, http://www.waveplane.com/

WCD 2000 Does Hydropower reduce Greenhouse Gas emissions?, World Commission on Dams, http://www.dams.org/news_events/press357.htm

WEC 2001 Survey of Energy Resources, The World Energy Council, London, http://www.worldenergy.org/documents/ser_sept2001.pdf

WEGA 2012 Sea for Life WEGA wave device, http://www.seaforlife.com/EN/FrameWEGAhow.html

Wello 2013 Wello Oy company website, http://www.wello.eu/

We@Sea 2010 Dutch Society for Nature and Environment/OMA offshore wind plan, http://www.we-at-sea.org/index.php?keuze=n&nummer=55

Williams A 2012 Run-of-the-river hydropower goes with the flow *Renew. Energy World*, Jan 31, http://www.renewableenergyworld.com/rea/news/article/2012/01/run-of-the-river-hydropower-goes-with-the-flow

Wind connection 2013 US Offshore grid proposal website, http://atlanticwindconnection.com

Windfloat 2012 Windfloat offshore turbine, REVE news coverage, http://www.evwind.es/2012/08/18/windfloat-project-ushers-in-a-new-era-of-offshore-wind-energy/21858/

Windstalk 2010 Atelier DNA design consultants, New York, http://atelierdna.com/?p=144

WNA 2012 World Nuclear Association website, and reactor data base, http://world-nuclear.org/NuclearDatabase/Default.aspx?id=27232

WWF 2012 A Study into the Economics of Gas and Offshore Wind, Cambridge Econometrics for WWF/Greenpeace, http://assets.wwf.org.uk/downloads/a_study_into_the_economics_of_gas_and_offshore_wind_nov2012.pdf

Yates N, Walkington I, Burrows R and Wolf J 2013 Appraising the extractable tidal energy resource of the UK's western coastal waters *Phil. Trans. R. Soc.* A **371** 20120181

Chapter 3

Heat

Renewable heat: bioenergy, solar thermal, geothermal

From the earliest times, human beings have used natural materials to keep warm and dry, and also learnt to burn wood and other naturally available materials, eventually including fossil fuels from underground. Wood and other forms of biomass, including animal dung, are still widely used for heating and cooking. Since 'fire' was humankind's first 'artificial' energy source, it perhaps should have come first in this book, but early use of biomass for burning was far from sustainable (land was often stripped bare), so it was not a renewable, or at least renewed, source. By contrast, modern use of biomass has involved improved efficiency of combustion, the use of new biomass sources and, crucially, backing up use with replanting. It has also involved the development of better energy conversion techniques, for example upgrading the age-old fermentation and charcoal-making processes. Biomass is also being used to make vehicle fuel.

In parallel, more use has been made of solar energy. Humankind has always used solar energy to keep warm by designing houses to capture and retain solar heat, and the design of houses for optimal 'passive solar' gain (as well as low energy loss) remains important. But the use of active solar collector devices for space and water heating is relatively new. The use of solar heat to make electricity is even more recent as is the direct use of light to make electricity (the latter being the topic of the next chapter).

The use of geothermal heat is also relatively new, although the Romans and others used hot water spas for bathing. Modern usage includes heating, but also using the heat from deep underground to produce electricity.

For the moment, although traditional biomass is still widely used, around the world heat is often provided by the direct combustion of fossil fuels, and electricity, generated from fossil fuel combustion, and more recently from nuclear reactions, is also used for heating. However, solar, modern biomass and biogas are increasingly being used for heating, and electricity produced from renewable energy sources is also feeding into the grid, and some of this is used for heating. In addition, new ways of converting electricity into heat more efficiently (heat pumps), and for using waste heat from fossil plants or

doi:10.1088/978-0-750-31040-6ch3

renewable sources (combined heat and power (CHP)/cogeneration) and storing heat (e.g. molten salt heat storage) have been developed. I will also be looking at them in this chapter on heat.

3.1 Bioenergy

The combustion of plant-based biomass provided humankind's first artificial and storable form of energy, foodstuffs aside, although we later learnt how to use oil from whales. We also leant how to convert biomass into charcoal, using this both as a fuel and to enrich soil. Some say we should do more of that, producing biochar to trap carbon dioxide as a form of 'black carbon' bio-sequestration (Fowles 2007), although there may be more cost-effective ways to sequester carbon (Elliott 2009, 2010).

Partial (constrained) combustion like this (i.e. pyrolysis) can also produce useful biofuels, as can the fermentation of some biomaterials (with alcohol production being the most obvious and long-standing human benefit!). Anaerobic digestion (AD) of putrescent materials is another (smellier) option producing biogas (methane). But full combustion still dominates the approach to biomass as an energy source. Whether for heat, electricity or use in vehicles, it is still burnt.

The use of biomass and wastes for electricity production is spreading. In 2011 the EU generated 123 TWh from these sources, the US 57 TWh and China 34 TWh. While some environmentalists worry about emissions, their main concern has been about the social and environmental impact of growing biofuels for vehicles, although some see energy and economic benefits (Worldwatch 2007). The debate on these issues has been long and sometimes heated. For example it has been claimed that growing biomass rather than food will lead to mass starvation, and that energy productivity is low and environmental impacts high. As a result, early assessments that biosources might provide large amounts of energy have been converted into much more cautious estimates: in 2012, the UK's Climate Change Committee suggested that the UK might only get 10% of its energy from biosources by 2050 (Elliott 2012).

It is usually argued that using biomass as an energy source is more or less carbon neutral, assuming new plantation replaces the biomass that is used. In that case, the carbon dioxide gas produced by biomass combustion is more or less balanced by that absorbed as biomass grows, less any generated by the use of fuels for harvesting, transport and processing. However a 2012 study for a group of UK environmental non-governmental organisations (NGOs) argued that, even with replanting, there will be significant excess net carbon dioxide added to the atmosphere since it will be emitted rapidly when biomass is burnt, but only absorbed slowly when new replacement plants start growing (RSPB *et al* 2012).

Clearly it is a dynamic system. There will be a delay after biomass has been harvested and burnt, before new plantings will start absorbing carbon dioxide. The report focused on trees and said the delay could be 'many decades'. It claimed that over a 20-year period, emissions from generating electricity using wood from conifer plantations were 1879 g kWh^{-1}, 80% more than using coal. It added 'Over a 40-year period emissions are lower because the trees have had longer to re-capture carbon, but even then biomass emissions would be 49% greater than coal power. Only after 100 years does electricity generation from conifer trees perform better than coal. And, regardless of the time period, it's never better than the current grid average and never meets (the Department of Energy and Climate Change) DECC's proposed maximum emission limit for Biomass.' The UK Biofuelwatch pressure group adopted a similar view (Biofuelwatch 2013).

This may be overstated in that, it has been claimed, there are other scenarios in which net carbon savings can be made (Forest Research 2012, BEG 2013). Even so, biomass is clearly not fully carbon neutral, especially if using trees (JRC 2013). However, the complete carbon life cycle is complex. Trees absorb CO_2 most rapidly when growing and slow down later in life. So although mature trees and their roots will still absorb some extra CO_2, they are less active and basically just carbon stores. Moreover they are not permanent carbon stores. They will eventually release it (as CO_2 or methane) when they die, rot or burn. So if carbon sequestration is the only issue, it might be best to grow biomass and chop it down regularly, before it is fully grown, and then grow more, while using it for fuel, so avoiding the use of fossil fuels and their emissions. An extreme version of this approach is represented by the development of genetically modified (GM) fast-growing plants like eucalyptus which, it is claimed, can grow 40% faster, thus speeding up CO_2 absorption and compensating for any short delay in absorption.

This approach highlights some of the problems with using trees for energy. Trees play key complex roles in ecosystems, not just as carbon absorbers, so it is important to understand what is happening when they are used for energy, not least in terms of water use. For example, fast-growing eucalyptus sucks water out of the soil rapidly, which could be disastrous in some locations, as has already been seen in Asia. Moreover, GM-enhanced biomass could also have other environmental and climate impacts, as well as, some fear, health implications (GJE 2013, Elliott 2011).

Issues relating to biodiversity, water use and changes in land use may turn out to be as important as the 'absorption delay' issue raised in the NGO report. Indeed, that report noted that the use of trees for combustion can divert timber from other uses, which means that more will have to be imported. This does seem to be unwise. So of course does the destruction of forests anywhere. That is why many environmentalists would prefer to stick with just forestry offcuts and trimmings, and farm wastes, along with domestic and commercial biowastes, as a source of biomass for energy use, and many would also prefer AD biogas production to direct combustion. Certainly co-firing biomass with coal, in old modified low efficiency coal plants, is not much of a step forward. We need to think about CHP and district heating, options I will be looking at later (section 3.4).

There is no question that the use of some types of biomass for energy are likely to be a poor choice. Some energy crops used for liquid biofuel production have very low calorific value, and mono-cultural plantations can compromise biodiversity, as well as requiring a lot of water and undermining local ecosystems. In addition to the use of biowastes, there may nevertheless be a role for some high-yield non-food energy crops on marginal land, and for less invasive approaches, such as short rotation coppicing.

Forests are a different matter. Forestry wastes may be useful but it seems clear that deforestation and unsustainable imports should be avoided and that attention should be given to other less damaging approaches to biomass sourcing and use, at various scales, including smaller-scale AD using food and farm wastes. Certainly mass burn of refuse, using its biomass content for energy production, has met with much opposition from local residents and environmentalists, fearful about toxic emissions (UKWIN 2013). Recycling can reduce waste, and pyrolysis and improved gasification techniques may help deal with what is left, but tight emission regulation is essential.

Perhaps the most contentious issues relate to growing biofuels for vehicle use. The independent Nuffield Council on Bioethics (NCB) has claimed that rapid expansion of biofuel production in the developing world has led to deforestation and displacement of indigenous people, the exploitation of workers, loss of wildlife and higher food prices. It has also contributed to poor harvests, commodity speculation and high oil prices, which raised the cost of fertilisers and transport. However, it also said that there was a clear need to replace liquid fossil fuels to limit climate change. If new biofuel technology can meet ethical conditions, it felt there was a duty to develop it. NCB concluded that an international certification scheme, like the Fairtrade scheme for food, was needed to guarantee that the production of biofuels met the five ethical conditions identified by the NCB: observing human rights, ensuring it was environmentally sustainable and reduced carbon emissions, and was fairly traded, with equitably distributed cost and benefits (NCB 2011).

The Food and Agriculture Organization (FAO) of the United Nations similarly claimed that bioenergy could be part of the solution to 'climate-smart' agricultural development, but only if its production was properly managed. In particular, large-scale liquid biofuel development may, they said, hinder the food security of smallholders and poor rural communities, and enhance climate change through greenhouse gas emissions caused by direct and indirect land use change. It was therefore crucial to develop bioenergy operations in ways that mitigate risks and harness benefits. Safely integrating both food and energy production addresses these issues by simultaneously reducing the risk of food insecurity and emissions, and Integrated Food–Energy Systems (FES) could, they claimed, achieve these goals on both small and large scales.

This may sound like wishful thinking, especially given the nature of the emergent industry, which some see as being rapaciously driven by the profitable vehicle fuel market. There have certainly been reports of poor working conditions in some biofuel plantations in Asia and elsewhere. But the FAO outlines ideas for how smallholder farmers and rural communities, as well as private businesses, could benefit from FES, offering a holistic view of the different types of energy that can be produced from agricultural operations, and how they can be aligned with current food production (FAO 2010).

The International Energy Agency (IEA) similarly seems convinced that, given proper controls, biofuels can play a major role. In its 2011 Roadmap, it said that they could supply 27% of global transport fuel by 2050, on a sustainable basis. The IEA said that 'while vehicle efficiency will be the most important and most cost-efficient way to reduce transport emissions, biofuels will still be needed to provide low-carbon fuel alternatives for planes, marine vessels and other heavy transport modes'. But it took the land-use issue seriously, drawing on an earlier IEA Biofuels report (see box 3.1).

Taking the land-use issue onboard, the IEA suggested that it would be possible to make use of 1 billion tonnes of residues/wastes, and 3 billion tonnes of high-yielding non-food energy crops, the so-called second-generation technologies, such as cellulosic ethanol. Even so, production would have to be supplemented with around 100m hectares of land, i.e. around 2% of total agricultural land, a three-fold increase compared with today. And the report admits that the 27% target would only be attainable if

Box 3.1. Facing up to the biomass land-use issue

A report entitled 'Bioenergy, Land Use Change and Climate Change Mitigation', published by IEA Bioenergy, notes that 'The effects of indirect land-use change are especially difficult to quantify and achieving a consensus on the extent of the impact is unlikely in the near future'.

However, it points out that 'bioenergy does not always entail land-use change. The use of post-consumer organic residues and by-products from the agricultural and forest industries does not cause land-use change if these materials are wastes, i.e. not utilised for alternative purposes. Food, fibre and bioenergy crops can be grown in integrated production systems, mitigating displacement effects and improving the productive use of land. Lignocellulosic feedstocks for bioenergy can decrease the pressure on prime cropping land. The targeting of marginal and degraded lands can mitigate land-use change associated with bioenergy expansion and also enhance carbon sequestration in soils and biomass. Stimulation of increased productivity in all forms of land use reduces the land-use change pressure.'

It concludes 'Bioenergy's contribution to climate change mitigation needs to reflect a balance between near-term targets and the long-term objective to hold the increase in global temperature below 2°C (Copenhagen Accord). While emissions from land-use change can be significant in some circumstances, the simple notion of land-use change emissions is not sufficient reason to exclude bioenergy from the list of worthwhile technologies for climate change mitigation. Sound bioenergy development requires simple and transparent criteria that can be applied in a robust and predictable way. Policy measures implemented to minimise the negative impacts of land-use change should be based on a holistic perspective recognising the multiple drivers and effects of land-use change' (IEA 2010).

lignocellulosic technologies were produced at an industrial scale within 10 years. There would also be a need for government support and research and development investment of more than $13 trillion over the next four decades and an international support programme. Nevertheless it claimed that 'biofuels would increase the total costs of transport fuels only by around one per cent over the next 40 years, and could lead to cost reductions over the same period'.

The IEA is far from complacent about the problems. It warned that the use of fossil energy during cultivation, transport and conversion of biomass to biofuel would have to be reduced, while direct or indirect land-use changes, such as converting forests to grow biofuel feedstocks which release large amounts of CO_2, would have to be avoided. The IEA said that it was important to impose sustainability standards for biofuels to prevent harmful impacts on land, food production and human rights. It suggested a land-use management strategy be imposed, along with a reduction in tariffs, to encourage trade and production of biofuels (IEA 2011a).

Are these proposals realistic? It ought to be possible, at least in theory, given the right regulatory framework, to avoid food–energy conflicts, but even with the best technology, there is still a risk that commercial pressures, locally and globally, for high added-value vehicle fuel production, will overwhelm any efforts at balance and integration. Biofuels could be the ultimate cash crop.

If we move away from high added-value products like biofuels for transport, there may be fewer problems. Biomass can also be used for heating and for generating electricity. Indeed many argue this makes more sense, since the final energy yields/acre using solid woody biomass are generally higher than for liquid biofuel production.

The Potsdam Institute for Climate Impact Research (PIK) has looked at the overall global potential for biomass and concluded that it could meet up to 20% of the world's energy demand in 2050, half of it from biomass plantations. But that would involve a substantial expansion of land use, by up to 30%, depending on the scenario, and irrigation water demand could double. In the PIK study, fields and pastures for food production were excluded, as were areas of untouched wilderness or high biodiversity, as well as forests and peatlands, which store large amounts of CO_2. But with second generation (non-food) energy crops, the bioenergy potential ranged from 25 to 175 EJ per year. The lower outcomes are for strong land-use restrictions and without irrigation; the higher outcomes assume few land-use restrictions and strong irrigation.

A middle scenario would result in about 100 EJ, while the world's energy consumption is estimated to double from today's 500 to 1000 EJ by 2050. In addition to biomass plantations and dedicated energy crops, PIK say that about the same amount of energy could be obtained from agricultural residues, and they see their expanded use as crucial for a sustainable future (Beringer *et al* 2011).

A study funded by the UK Energy Research Centre (UKERC) came to similar conclusions, at least on the benefits of using non-dedicated land, in the UK context. It looked at the potential of planting short rotation coppice (SRC) using fast-growing poplar and willow, in England, taking into account social, economic and environmental constraints. It concluded that planting SRC energy crops on England's unused agricultural land could produce enough biomass to meet renewable energy targets without disrupting food production or the environment (UKERC 2011).

UKERC said that new technology would enable biofuels for energy use to be made from lignocellulosic crops (e.g. SRC willow and poplar) which, unlike current cellulosic crops (typically derived from food crops such as wheat and maize) could grow on poor-quality agricultural land. While the results suggest that over 39% of land in England cannot be planted with SRC due to agronomic or legislative restrictions, marginal land is realistically available to produce 7.5 m tons of biomass. This would be enough to meet around 4% of current UK electricity demand and 1% of energy demand. The south west and north west were seen as having the potential to produce over one third of this, given their large areas of poor-grade land.

Not everyone will agree that, even with new types of crop, biomass can be much of an option, especially if fuels for vehicles are included, but the potential global resource is very large, depending on what is included as available biomass, ranging from around 100 EJ (about 1/5th of current global primary energy supply) up to 600 EJ (slightly more than current global primary energy supply) assuming wide use of energy crops covering over 10% of global land area (Slade 2011).

Extreme scenarios like that are probably out of the question, but some radical technical fixes are emerging that might improve the situation for biofuels and/or biomass use. Perhaps the most obvious idea is to use biomass differently. Certainly biogas production via AD, e.g. of biowastes, is widely seen as a good idea, especially if coupled with using the gas grid for delivery. It can then be used for heating or for electricity generation. Biogas can also be used in car engines. For example, it is claimed that methane, through AD, requires only about a quarter of the land area ethanol requires, and is a far more efficient fuel than ethanol, easily used in cars and trucks (CNG 2013).

Another idea is to produce hydrogen by thermo-chemical processing of biomass/wastes, as proposed by Karl-Heinz Tetzlaff. The hydrogen can be used for heating, possibly admixed with methane and delivered via the gas main, in fuel cells for electricity production, as a vehicle fuel or to make syngas/ammonia (Tetzlaff 2010). I will be looking at more examples in chapter 5, since this type of approach offers options for system integration. I look at fuel cells in chapter 4.

There are some efficiency penalties involved with these various conversions and, even if biomass is being used more efficiently, the land-use issue remains. However there are more radical approaches, which might reduce the latter. For example, 'Breaking the Biomass Bottleneck', a report from Concito, a green 'think tank' in Denmark, suggested that biomass could be upgraded by hydrogenation, using hydrogen produced by the electrolysis of water, powered by excess electricity from renewables like wind. Then there would be less land needed per kWh of final energy from the bioproduct.

The report claimed that you can react biomass with hydrogen 'to produce hydrocarbons of much higher energy content and energy density than the original biomass. Moreover, using the biomass and the biogenic carbon from hydrogenation in central applications like heat and power, it is possible to collect the CO_2 from the biomass and further recover and recycle it in a process here called carbon capture and recycling (CCR). This will further multiply the use of the biogenic carbon from the biomass. Overall, upgrading and recycling biogenic carbon by hydrogenation and CCR, can approximately quintuple our biomass potential for providing storable and high-density fuels and carbon feedstock compared to the presently applied technologies for converting biomass to fuels and feedstock' (Wenzel 2010).

This sounds like getting something for nothing, but in reality it does not invalidate the laws of thermodynamics. Although the report noted that, even with electrolysis losses, 1 joule of wind can save 1 joule of biomass, by upgrading it, it added 'the total energy content of the biomass and the hydrogen is, of course, greater than that of the fuels on the output side. If, therefore, hydrogen is sufficiently good for the demanded energy services in question, there is no sense in taking a detour of producing the carbon-based fuels from the hydrogen. The conversion from hydrogen to carbon fuels as energy carrier is only justified by the inherent differences in the properties and qualities of the two'.

So it is end-use utility that matters, especially as it costs more, given the efficiency losses. Even so, the report claims that the result is a valuable green fuel, which can be stored easily and be used to help balance variable wind and other renewables, while needing less biomass and less land. Moreover, if the biomass used is replaced sustainably, and if the CO_2 produced when the fuel is burnt is captured, then it is overall net carbon *negative*, although CO_2 capture would not be possible for transport uses.

A review by the Fraunhofer Institute in Germany illustrated that this idea can include a range of renewable inputs, not just biomass, but with biogas and other bioproducts as the final outputs (Sterner 2010). I will be looking at that idea in chapter 5. It might be expensive, but adding carbon capture and storage (CCS), that is storing captured carbon dioxide emissions in wells/strata underground, could give it an edge over other energy options.

I will be looking at CCS shortly, but first, another very ambitious approach to producing biomass, while reducing the land-use issue, involves using algae or other biomass grown in low land-value desert areas, possibly coupled with CCS. It has been argued that, if algae is grown at the yields that the IEA Task Force Bio-energy said was credible, then a land area the size of 24% of Australia (in practice spread around Earth's deserts) would produce 90 000 TWh yr^{-1}, which is nearly equivalent to the current global final electricity demand of 98 000 TWh yr^{-1}. Moreover if that algae/biomass is then used in CCS schemes then it could be carbon-negative (GCCSI 2010).

One way to do this might be by growing algae or other biomass, as well as food, in seawater greenhouses, cheap poly-tunnel constructions in desert areas fed with seawater. A commercial scheme is up and running in Australia and others have been built in Tenerife, Oman and Abu Dhabi (SWG 2013). In parallel, the Sahara Forest group is building a major project in Qatar, which also includes solar technologies to desalinate water, and another is planned in Jordan (Sahara Forest 2013). Of course there are many uncertainties in relation to, for example, costs and impacts on fragile desert ecosystems, but the potential energy resource is very large.

There is another idea: growing algae at sea or on lakes. Researchers at the University of Texas have suggested that 'the surface area of water required to grow enough algae to fully replace petroleum motor diesel in the US is slightly less than that of the Great Lakes, which cover 24.4×10^6 hectares' (University of Texas 2012). That either sounds promising or terrifying, depending on your perspective.

I have dwelt on biofuel impacts and issues at some length, since many 'greens' are worried about the use of biofuels to keep the cars (and planes) going for a range of reasons. There are clearly limits to how much we can rely on biomass for transport fuel and also for heating and electricity production, but technology may help raise some of those limits and also provide more utility from biomass. For example, it does not help with vehicles but, as noted above, there is some enthusiasm for using carbon capture technology with biomass combustion, since 'biomass energy with carbon capture' (BECC) makes biomass combustion not just carbon neutral over time (allowing for a delay in re-absorption), but actually carbon *negative*.

Many environmentalists do not favour the use of CCS for fossil fuels, since they fear that it will just allow for the continued use of polluting energy sources, delay the development of renewables, and will not, in any case, be effective, safe or economically viable. Certainly there could be problems (Zoback and Gorelick 2012). Nevertheless some see its use with biomass as possibly a good option. However, not all are agreed. A Biofuelwatch report says that BECC, like CCS in general, will not be economic, or (given delayed re-absorption) carbon neutral, much less carbon negative, and it worries about 'underground land-grabs' for storage space.

Given their concerns about the environmental implications of biofuel production, they also worry about ethanol production involving fermentation, which results in a pure stream of CO_2 that can be directly captured, and biodiesel production using the Fischer–Tropsch method, which involves production of syngas as an intermediary step, and offers opportunity for CO_2 capture. Overall, even with high-yield biomass crops, they do not think there will be enough land to allow for significant *sustainable* biomass production, so BECC is unlikely in any case to be much of an option (Biofuelwatch 2012).

The debate over biomass and biofuels continues. The fact that the term biomass covers both biomaterials for heat and electricity, and also fuels for vehicles, may confuse what could be separate issues. Although land-use/biodiversity/eco-impact issues are common to both types of biomass use, the high economic value of vehicle fuel arguably makes that much more of a threat. By contrast, the potential for generating heat and also electricity, using solid biomass from SRC, or biomethane from waste streams, is large, and these options may not be so problematic, as long as there is tight regulation.

Biomass has had a long and sometimes turbulent history as an energy source. People have often fought over getting or retaining access to it, or to profit from or control its use. Some of this has been part of the struggle over land rights in general, for food production. But certainly the use, and overuse, of land has led to major environmental problems, including widespread deforestation. So the modern debate may not be anything new.

In this review I have hardly touched on the large range of projects and ideas for bioenergy conversion and processing. As I have indicated, most involve converting biomass into fuels or using heat from biomass or biogas combustion to make other forms of energy. But there are new ways of doing this, including CHP which, as I mentioned briefly in chapter 2, uses the waste heat inevitably produced by electricity generation. I will be looking at that in detail later. It has wide application, for all energy sources, at least doubling overall energy conversion efficiency. For now just note that, as table 2.2 in chapter 2 indicated, in the UK context, biomass energy conversion to electricity and heat, via CHP, has load factors of 80–90%; it is highly efficient.

There may be an environmental price to pay for high conversion efficiency. For example combustion of biomass, and wastes especially, can produce unwanted toxic emissions, unless carefully controlled. I have described how somewhat brutal approaches like combustion might be at least augmented by more gentle, slower processes like AD. But the basic problem is that getting at the energy content of biomass in environmentally sound ways is hard.

There may be new ways. I will end this section with a perhaps quixotic example, a way to prise the sugars (and hence energy) in wood out of the tough structures in which they are encased. The gribble, a 2 mm creature, like a very small woodlouse that lives in the sea, can digest wood. Researchers are looking to harness the power of the gribble's enzymes to release the sugars from wood that can then be fermented to make biofuels. They say the major advantage is that 'waste materials instead of food crops can be used to make fuels, delivering a double bonus in not competing with land for food production as well as utilising unused materials from timber and agricultural industries' (BBSRC 2012).

3.2 Solar heat

Biomass use for heating has had a very long history. By contrast, the direct use of solar energy for heating is a relatively recent development, although in antiquity there were a few interesting solar-powered devices, e.g. for making jets of steam. The Victorians also developed solar-driven machines and heating systems and there are some fascinating early examples of solar energy use (Butti and Perlin 1980). But now the use of modern domestic solar heat collectors for space and/or water heating is very widespread, most obviously in hot climates (in Greece for example), but also elsewhere (Austria is one of the leaders). Indeed when the heating season is long, as in cold northerly locations, the value of any solar heat that is available is much higher than it would be on the Equator.

The basic technology for domestic use is relatively simple. Blackened radiator-like units under a glass cover, or small evacuated glass cylinders with light focused on a central axial tube, absorb solar heat and feed heated water to a boiler for use in central heating systems. Large units are also commonly used for hotels, swimming pools and so on. Moving upscale there are now some very large partly solar-fired community-wide district heating networks in Denmark and elsewhere. I will be looking at examples later. In addition, rather neatly, as we shall also see later, solar energy can be used for cooling.

Solar heating systems are environmentally benign, the only significant impacts being a small amount of visual intrusion (if on roofs) and the need for water for surface cleaning. As with some standard heating and ventilation systems, there can be problems with Legionnaires' Disease if they are not installed and maintained properly.

Globally, there is nearly as much solar heat capacity in place as wind energy capacity, and it is expanding. According to the *Solar Heating and Cooling Roadmap* of IEA it could account for around one-sixth of total global low-temperature heating and cooling needs by 2050. This, it says, would eliminate some 800 Mt of CO_2 emissions per year. This is more than Germany's total CO_2 emissions in 2009. By 2010 there was 195 GW(th) installed globally (118 GW of it in China), rising to 245 GW(th) by 2011.

The IEA roadmap claims that, if governments and industry took concerted action, solar energy could annually produce more than 16% of total final energy use for low-temperature heat and nearly 17% for cooling by around 2050, a 25-fold increase in absolute terms of solar heating and cooling (IEA 2012).

The IEA notes that, in addition to replacing fossil fuels that are directly burned to produce heat, solar heating technologies can also replace electricity used for heating. This it said would be especially welcome in countries without gas infrastructure and lacking alternative heating fuels. South Africa is cited as an example of a country that would benefit, as electric water heating currently accounts for a third of average household (coal-based) electricity consumption there.

The report also notes that solar thermal cooling technology, in which the Sun's heat is used to power thermally driven absorption chillers or evaporation devices to cool air, can reduce the burden on electric grids at times of peak cooling demand, by fully or partially replacing conventional electrically powered air conditioners in buildings. As climate change impacts, cooling is going to become a major issue around the world, not just in currently hot climates, and direct solar cooling has obvious attractions, as long as added-on technology does not replace intelligent design of buildings to limit overheating. Over the centuries, so-called 'vernacular designs' have emerged in hot climates around the world to keep buildings cool but, in many cases, sadly, they are being replaced by modern, often high-rise buildings, which need energy-using cooling systems to make them habitable (Sahakian 2011).

The IEA Roadmap also stresses the scope for expanding use of these technologies in industry. Often overlooked is several industry sectors' significant energy demand for low- and medium-temperature heat in such processes as washing, drying agricultural products, pasteurisation and cooking. It says those industrial processes offer enormous potential for solar heating technologies, which could supply up to 20% of total global industrial demand for low-temperature heat by 2050.

Given the variable availability of solar energy, a key area for development and support is storage. In theory, individual domestic solar collector systems could have extra thermal storage (over and above that provided by their normal boiler). However, there are some significant advantages to operating at larger scale. For example, on top of the savings from the bulk buying and installation of large numbers of units, there are efficiency gains from the use of large heat stores; the ratio of their outer surface area to the contained volume decreases with size, so the rate of heat loss decreases. In addition, instead of having to match the heat demand patterns of an individual household, a large store can serve many houses, so that the often different individual demand patterns are averaged out, leading to higher system efficiency and lower costs.

This approach is sometimes called 'grouped solar', with individual housing blocks or terraces sharing a large solar array and large heat store. But to do this on a larger scale, for a complete community or urban area, there has to be a district heating pipe network, and that adds to the capital cost, although once installed the running cost is low, and the solar-fed heat is free, with costs falling for larger systems. Indeed some say that, in many locations, solar only makes sense with district heating (Grydehoj and Ulbjerg 2005).

There are many solar heat collector projects around the EU linked to district heating networks, backed up by large heat stores, some of them being inter-seasonal stores, storing summer heat for winter use. The largest so far is Marstal's 13.5 MW solar array with linked heat store, in Denmark (Marstal 2013). Some have large well-insulated water tanks or lined pits, or engineered thermal masses. Underground thermal energy storage (UTES), with excess heat simply stored in the ground in the summer to be extracted in the winter, may be cheaper. Some systems use deep vertical boreholes, such as the Drake Landing inter-seasonal solar heat storage system in Canada, where of course winters are very cold (Drake 2013).

I will be looking at energy storage more generally in chapter 5, with heat storage being one option. At present, as I will be describing, the major focus for storage tends to be for electricity, but whatever the initial impetus for developing storage, solar heat technology could certainly benefit, and heat storage technology at various scales is becoming a new focus of attention, as is district heating; see box 3.2.

The use of domestic/community scale solar heating is of course only one part of the process of dealing with energy use in buildings. There are many energy saving and building design issues which also have to be addressed. I will look at some in chapter 5, and at the use of heat pumps later in this chapter. At the very least buildings should be designed to reduce heat losses and to maximise passive solar gain, for example through well placed windows/glazing (WBDG 2012).

3.2.1 Concentrating solar power

Solar space and water heating are clearly important, and as table 2.2 in the previous chapter indicated, load factors can be reasonable, at 50% in the UK context, and of course higher in sunnier locations. However, now I want to look at another role for solar

Box 3.2. Solar heat storage and district heating

It is clear that the cost of solar falls significantly with the scale of the system (Steffersen 2007). As a result large solar arrays linked to district heating networks are spreading.

Denmark is the leader in solar-fed district heating (DH), with 85 MW(th) in place, some with solar heat stores (Preheat 2007). It has ambitious plans for expansion. Around 60% of Denmark's domestic/commercial heat is supplied by DH (some fossil fired, some also fired by biomass) but by 2015 it wants 3% of this to be solar fed, rising to 10% in 2030 and 40% by 2050.

Austria is also moving ahead in the solar DH field. There is a district heating network in Graz, with 6.5 MW(th) of solar input and more are planned. Germany has installed nine research and demonstration solar arrays linked to district heating networks since 1996, including some with *inter-seasonal* heat stores. Depending on their size, they can meet 40–70% of the annual heating needs of a building. District heating of any sort is rare in the UK, but the Centre for Alternative Technology in Wales pioneered a site-wide heating network fed in part from solar, with a large pit-type hot water heat store, backed up by a woodchip boiler. UK company ICAX is developing similar ideas with commercial-scale solar-fed inter-seasonal heat stores (ICAX 2013).

This is not just a European phenomenon. The Drake Landing project in Canada has 52 houses with solar collectors and 144 × 35 m deep heat storage boreholes (Drake 2013). But Europe leads, with ambitious plans for expansion (SDH 2013).

heat: electricity production. As with heat from biomass, solar heat can also be used for electricity generation, so in some ways you might see both as falling in the area covered by the previous chapter. But in both cases the emphasis, in terms of the primary energy source and conversion process involved, is on heat, unlike with hydro, wind, wave or tidal energy conversion, where no heat is used or produced (other than incidentally).

Solar thermal electricity production (as distinct from photovoltaic electricity production, which I look at in chapter 4) involves boiling water to raise steam (or using other fluids/vapours) to drive turbines, much as with conventional thermal power plants, whether fossil fuel or nuclear fired. To get high enough temperatures, with concentrating solar power (CSP) plants, sunlight is focused on a small central receiver with a boiler, using large numbers of reflecting mirrors tracking the Sun, or use is made of single parabolic dishes, or reflective material coated cylindrical troughs with heat collector pipes at the focal points. For maximum sunlight and improved economics, CSP units are usually located in sunny desert areas, with some of the first being in the Mojave desert in the US, and in Spain. Subsequently projects have emerged in North Africa and the Middle East; see box 3.3.

The energy potential for CSP is huge: there is a lot of desert! 1000 GW or more could in theory eventually be installed globally. Note that some call it 'concentrated' solar, not 'concentrating', but whatever the label, the technology exists, with over 2 GW in place around the world, and its economics and performance are improving. IRENA has suggested that capital costs could fall by up to 50% by 2020 (IRENA 2013).

Initially most CSP plants have been designed to run as hybrid plants, along with gas-fired turbines, with the solar contribution as ancillary. The gas input allows for continued

Box 3.3. CSP in the MENA region and beyond

There are already several CSP units working in the Middle East–North Africa (MENA) region, and others are planned. Egypt has a 150 MW hybrid CSP/gas plant running just outside of Cairo. Its National Plan for 2018–22 has 2550 MW of CSP. Abu Dhabi has 100 MW of hybrid CSP in place, while Morocco has a 470 MW hybrid solar/gas fired unit, with 22 MW of CSP, and Algeria has a 25 MW unit, backed by a 130 MW gas-fired plant. Tunisia's 'TuNur' CSP project should ultimately have 2 GW capacity. There are also projects in Jordan, Israel and elsewhere in the region and some ambitious plans: the UAE is looking to have 1 GW of CSP/PV, Qatar is to invest up to $20bn in a 1.8 GW solar plant scheduled for 2014, and Saudi Arabia plans to have 41 GW of PV/CSP solar (25 GW CSP, 16 GW PV) by 2032, via a $109bn programme. Meanwhile, CSP continues to develop in the USA and Spain, while China, India and South Africa have been looking at possible CSP projects (CSP Today 2013).

A global review in 2009 claimed that, on the basis of moderate assumptions for future market development, 'the world would have a combined solar power capacity of over 830 GW by 2050, with annual deployments of 41 GW. This would represent 3.0 to 3.6% of global demand in 2030 and 8.5 to 11.8% in 2050'. Moreover 'under an advanced industry development scenario, with high levels of energy efficiency, CSP could meet up to 7% of the world's projected power needs in 2030 and a full quarter by 2050' (Solar PACES *et al* 2009). Since then, the global recession has had an impact on CSP growth, especially in Spain, hard hit by the Euro crisis, but projects in the MENA region are still progressing, and projects are planned in Australia, India and Chile.

operation at night, using the same turbines. But increasingly, use is being made of molten salt heat stores, taking some daytime heat, for 24/7 running. Typically they use a mixture of 60% sodium nitrate and 40% potassium nitrate and can store heat overnight with low losses, and release it at night for continued steam raising (Way 2008, EERE 2013).

There are some question marks about the local environmental impact of CSP. Although deserts are often seen as fairly sterile environments, the ecosystem that does exist is very fragile and easily disrupted. For example, there have been objections to some large CSP projects in California on the basis of the potential impacts on desert tortoises and other local wildlife (Solar Done Right 2010). Perhaps more significant is the fact that CSP plants have to be cooled and, although air cooling is possible, it is less efficient, making power production more costly, so that water cooling is preferable. However, water is one thing deserts do not have, and what is there is needed by the local ecosystem and by any people in the region (NRDC 2009).

Fortunately, in the North African context, CSP projects can be near enough to the sea to allow for water to be piped in, long distance if necessary. Piping in seawater is not seen as being very expensive. That is important since, as well as needing cooling for thermal plant operation, the CSP mirror surfaces will need regular washing to clean dust off them. An alternative may be electrostatic cleaning techniques of the type developed for use with PV cells on lunar and Mars landers (Jalbuena 2010).

In addition to providing electricity for local use (including possibly desalination of sea water), some enthusiasm has been expressed for the idea of exporting it long distance using high voltage direct current (HVDC) supergrids, for example, with an undersea grid link from North Africa across to Europe. HVDC transmission losses are put at around 2–3% per 1000 km. So long-distance export seems credible, although clearly that will involve some disruption for the land link part with new grids being installed.

The German (E.ON) backed Desertec Initiative has made most of the running, with talk of a $400bn programme to link the EU to CSP projects in the MENA region with HVDC links (Desertec 2013). Although at present it is mainly just a concept, some independent CSP projects are being backed. There is a parallel French (EDF) backed Transgreen programme (subsequently it seems revamped as Medgrid) focusing more on the grid side with, for example, a plan for a link-up to Morocco (Medgrid 2013).

These proposals tie into the wider 'Solar Med' programme promoted by the fledging Mediterranean Union (now apparently recast as a Euro-Mediterranean Partnership), which included plans for a 'Mediterranean Ring' grid system, linked to 20 GW of renewables with, by 2020, around 6 GW of wind, 5.5 GW of CSP and nearly 1 GW of PV solar in North Africa and the Middle East (Europa 2013, MSP 2013).

Proposals like this raise some interesting practical questions. For example, would the French HVDC supergrid link to North Africa have to go across Spain to Gibraltar, or could it go undersea directly, off the Spanish coast? That would be much more expensive, 1200 km of marine cable at maybe €1m/km, rather than a short stretch across to Morocco. The trans-Spain option does also allow Spanish wind power to be fed in, but the undersea line avoids land use and local or regional or indeed national and trans-national political conflicts. Spain has recently had a major a battle getting a HVDC grid link across the Pyrenees to France. It had to be put underground at large extra cost.

There are also some wider strategic issues. There is the risk of a neo-colonial resource grab, with investors rushing to get sites and then seeking to buy/sell the electricity at favourable rates. It will be vitally important to negotiate fair trading arrangements. However, there is also a need to consider what is wanted locally. While the EU media has often waxed lyrical about the concept of energy from the desert, at one point some countries likely to be hosts to CSP projects to supply the supergrid evidently felt they had not been sufficiently consulted (Brand 2010). They may have their own differing views. Energy demand is rising rapidly in the region and there may not be much enthusiasm for exporting any CSP power, in preference to using it locally, e.g. for desalination. The Desertec Industrial Initiatives' 2050 projections certainly suggested that energy demand in some MENA countries could be high by then (DII 2012). That said, the Desertec plan is based on exporting only around 15% of the CSP energy to the EU, so there should be plenty for local use. But then again, CSP may not appeal locally. In the MENA region, electricity prices are often heavily subsidised (e.g. in Egypt) for social policy reasons, and that would be hard to change. So, initially at least, the relatively high price of CSP electricity might mean that it could only realistically be sold abroad.

Supergrid links raise many other issues, not least their vulnerability to hostile attack. One response to that is that the grid system could be a multiple line network, with redundancy built in, like the internet, so the temporary loss of one line would not matter. Attacks on CSP plants are a risk, but that is true of all energy facilities, as witness the 2013 attack on a BP oil facility in Algeria (and, of course, the World War II RAF

Dambuster raids!). However, given that CSP plants would be earning export income, there would be a strong national incentive to protect the sites (Stegen 2012). Similarly that would make it unlikely that solar producers in the south would seek to hold the north to ransom, by cutting off supplies. Unlike oil, solar energy cannot be stored for long.

For the moment, the main emphasis in terms of supergrids is the North Sea, to tie in offshore wind projects and aid cross-national boundary electricity trading and grid balancing. But it seems unlikely that the huge solar resource in the south will be ignored for long. Some argue that the north should not import green energy, but should focus on its own resources. Certainly the EU should not use CSP as an excuse not to develop its own extensive renewable resources as fast as possible, large and small, locally and nationally. Indeed the potential for large- and small-scale renewables is so large within the EU that some say it should not need imports from CSP or anything else. Nevertheless, a fully integrated supergrid system, including links to offshore wind, wave and tidal energy in the north west of the EU, as well as to CSP in desert areas, could offer benefits to all, in terms of helping to balance variations in the availability of renewable energy around the entire region. And locally, CSP could offer electricity, fresh water, jobs and income.

Moreover, as far as the planet is concerned, it does not matter where the projects are located, as long as they reduce/avoid emissions. If the best sites for solar are in desert areas, so be it. But politics and economics intervene. With projects also being considered in Asia (e.g. a supergrid link to proposed CSP projects in Mongolia), the CSP/supergrid concept opens up a range of new and old geopolitical and development issues, not least who gets the energy, at what cost, and who gets the profits. I will be looking at some of these wider issues in later chapters.

CSP is not the only new idea for using solar heat to make electricity. One even larger scale idea is to have vast solar-heated glass greenhouses surrounding a chimney up which the solar-heated air passes by convection, turning an internal air turbine near the top. The so-called 'solar chimneys' or 'solar updraft towers' would have to be very tall to get sufficient acceleration of the updraft. Nevertheless, a small prototype was built in Spain in the 1980s and full-scale projects (with chimneys of 500 m or more in height) have been mooted in Australia and elsewhere (Schlaich *et al* 2005). Less dramatically, but perhaps with wider potential, solar heat can be collected and stored in ponds or lakes, sometimes using covering membranes or salt water gradients to limit heat loss, with the heat extracted for electricity generation. Solar ponds have been built and run in Australia, India and the USA, some with an eye to desalination of seawater (Green Trust 2013).

A very different approach to heat collection and electricity generation is represented by ocean thermal energy conversion (OTEC), which makes use of the temperature differential between surface waters and water at depth in the sea. This would be large-scale technology, involving huge floating devices with pipes reaching deep down into the ocean and large heat engines working on the temperature difference. It has been suggested that the global energy resource could be up to 100 TW, but it is hard to see how the electricity generated, presumably far offshore, would be delivered to users on land. It could perhaps be converted at source to hydrogen to be tanked ashore, for shore-based generation plants or vehicle use (OTEC 2013). As with solar ponds and solar chimneys, OTEC is also only relevant to areas of the world where there is high insolation. For OTEC, that mainly means the Pacific. Moreover there could be ecological issues.

Another approach is to use focused solar heat to make hydrogen by thermal dissociation of water molecules. This needs very high temperatures and the process is not very efficient. A more indirect approach, as in the Solar Gas project in Australia, is to use focused solar heat to convert a mixture of methane (natural gas) water and/or CO_2 into new higher energy value synfuels. CSIRO claim that this offers an extra 25% energy gain (CSIRO 2013). A US system claims to get a fuel use reduction of around 20% by using heat from a parabolic focused solar dish to convert fossil gas to syngas, as a fuel for a gas turbine (Science Daily 2013). Gasified biomass could also be used as a feedstock for hybrid solar-gas systems like this, producing hot gases for gas turbines. Going beyond that, US company Joule says it has developed a solar energy driven conversion process for making waste CO_2 into the key components of sulfur-free petrol and jet fuel, without the need for biomass feedstock. It is claimed that it needs 'only adequate sunlight, access to waste CO_2 and non-potable water' (Alspach 2013).

Novel uses of solar heat like this cross classification boundaries, taking us back to biomass, as a joint input (with solar) to an electricity generation system, but also to green synfuels as a possible output. As I will be exploring in the next chapter, the thermal part of light can also be used directly to make electricity in solar cells, so that might have been included in this chapter.

However, beyond classification issues, a key point that emerges in this context is that the solar heat resource can be used flexibly for a range of applications and much work is being done on hybrid systems. For example, there is a €22m EU-backed 1 MW 'Multi-Purpose Applications by Thermodynamic Solar' project in Egypt using CSP and biogas to supply electricity and desalinate water (Almohsen 2012).

Moreover, in addition to electricity supply, as I have indicated, solar can be used directly for heating on a small local scale for houses or communities, although storage is important and that can make solar heating expensive, even at a large scale. Nevertheless, low-grade solar heat can be upgraded by the use of a heat pump for domestic use (Minus7 2013, Infinity 2013), another indication of the flexible use of solar heat. I will look at heat pumps later, after I review the use of geothermal heat. Like solar energy, geothermal energy can be used for either heating or for electricity generation, but also possibly both at the same time although, at least in terms of electricity generation, with geothermal, small-scale operation is less of an option.

3.3 Geothermal heat and power

Hot water from geothermal aquifers has been widely used for heating. In 2010, around 50 GW(th) was in use globally, including ground-sourced heat pumps. In addition there was about 11 GW(e) of geothermal electricity generation capacity globally (Lund *et al* 2010). The geothermal resource relies on the radioactive decay of some isotopes deep underground, so it is really a 'natural nuclear' source. As such it is renewable in the sense that this decay process will be very long lived. Nevertheless, the local heat gradient that is established will be reduced by heat extraction so, over time, geothermal projects may gradually lose power. The heat gradient will be re-established in time, if extraction is halted, but a new well will have to be established somewhere else for the duration.

These problems are unlikely to affect shallow aquifer projects, only the deeper wells aiming for higher temperatures, which is what is needed to generate electricity. Depths of 3000 to 10 000 m can now be reached. Water is pumped down to be heated by the hot rocks to around 200°C, and fed back to the surface to drive turbines.

The best/easiest sites are often in granitic areas. Even so, the deep well resource is very large. Indeed, very deep down, there is heat everywhere, and the theoretical resource is larger than could ever be needed. In practice even drilling wells a few kilometres down is hard, expensive and risky, and there may be environmental implications.

For deep geothermal electricity production, one approach is to drill two wells close by and then pump fluid down one under high pressure to cause rock fracturing at the bottom, thus creating a complex series of paths through to the other well. A heat collector is thus created. The technique has some similarities with the 'fracking' technique used for shale gas wells, except it only seeks to create a relatively small nest of fractures deep down. Even so, as with shale gas fracking, that can trigger micro-earthquakes in local faults, as happened at a geothermal well drilling project in Switzerland in 2006, when water was injected at high pressure into to 5 km deep borehole. A shock measuring 3.4 on the Richter scale was detected, which caused local alarm, but evidently no injuries or serious damage, although further work was halted. There was a 3.1 scale tremor later. Clearly it is an issue, but possibly not a major one (Rein 2011).

In operation, cold water is pumped down one well and it comes out hot up the other. It may be contaminated with materials from underground including (toxic) hydrogen sulfide gas, which may escape. There may also be some carbon dioxide gas release. (Veal 2009). However, in order to limit the amount of fresh water that has to be injected, closed-loop systems are increasingly used, with cleaned-up water being reinjected.

The economics depend crucially on the site and its productivity. The key requirement in forming the well heat-exchangers is to ensure that the path through is not too easy (or else the water passes through quickly and does not pick up much heat) or too hard (in which case the water only dribbles through and needs a lot of pump pressure to get an output). Well productivity is thus hard to predict in advance, and several wells may have to be drilled to get it right. That is expensive, although it has to be remembered that this is normal for oil companies. They do not give up if an exploratory well fails to deliver. They try again, since the payoff for success is high, and that is also true for geothermal. Once running, geothermal energy projects, unlike solar wind, wave or tidal projects, can deliver continuous energy output. Moreover, new 'enhanced geothermal' technology may make geothermal even more attractive. For example, usually geothermal plants require sources of 182°C or more, which are hard to find and costly to develop. By contrast, newly developed low-temperature geothermal technology makes it possible to use resources at less than 150°C, or even 74°C (Chandrasekharam 2008).

In terms of new plants, the USA leads, with over 120 new projects under development, around 4 GW of new capacity. Google.org recently put $5.4m into enhanced 'hot rock' geothermal systems, supporting three new projects in the USA, and President Obama allocated $350m to geothermal work under the economic stimulus funding.

The resource potential is seen as being very large. A 2006 MIT study suggested that the US could have at least 100 GW(e) of geothermal capacity by 2050 (MIT 2006). Many other countries also have large resources. Iceland is a leading user, but the Philippines, which generates 23% of its electricity from geothermal energy, is the world's second biggest producer after the USA. It aims to increase its installed geothermal capacity by over 60%, to 3130 MW(e). Indonesia, the world's third largest producer, plans to have 6870 MW(e) of new geothermal capacity over the next 10 years—equal to nearly 30% of its current electricity generating capacity from all sources. Kenya has announced a very ambitious plan to install 1.7 GW(e) of geothermal capacity within 10 years. There are many projects in operation or being developed elsewhere around the world (IGA 2013).

A 2.9 MW(e) plant is operating commercially in Landau, western Germany and many more are planned. Norway is also looking at geothermal. Japan, which already uses geothermal heat widely, is hoping to have up to 2 GW of electricity generation capacity by the 2020s. In the UK, geothermal could supply 20% of UK electricity from around 9.5 GW of installed capacity, according to a 2012 report by consultants SKM. It added that projects with a 25-year lifetime could in theory also support 100 GW of heat supply capacity, meeting all UK space heating needs (SKM 2012). See box 3.4.

As already noted, the global potential for geothermal is large, but it is helpful to put that in perspective. The global average is a 25°C rise per km down. Martin Culshaw of the UK's Geological Society's engineering group, has said: 'Cooling one cubic kilometre of rock by one degree provides the equivalent energy of 70 000 tonnes of coal. This has the potential of equalling the nuclear industry in providing 10–20% of Europe's energy.' Moreover, new drilling techniques may help make it more viable, like the laser system being developed in the USA, useful if you are drilling through granite (Hecht 2012).

Geothermal may thus represent a very significant new energy option.

Box 3.4. Geothermal energy in the UK

There is an aquifer hot water system operating in Southampton, linked to a DH network. Some pioneering deep geothermal 'hot dry rock' work was also done in Cornwall between 1976 and 1991, but it was halted on cost grounds.

Until recently the UK government's commitment to geothermal has been relatively limited, although there are now some projects underway. EGS Energy is developing a 3 MW(e) demonstration project on a site at the Eden Project in Cornwall. It was awarded £2m via the Deep Geothermal Challenge Fund (EGS 2013).

Geothermal Engineering Ltd is also planning a plant near Redruth in Cornwall, with a well reaching 4.5 km below ground level to access rocks at temperatures of around 200°C. This will provide up to 55 MW of renewable heat and 10 MW of electricity (GEL 2013).

Three geothermal projects run by Keele, Newcastle and Durham Universities and Cofely District Energy in Southampton are sharing £1.1m from the Government's Deep Geothermal Challenge Fund. There is also a 10 MW(th) project planned in Manchester, backed by E.ON (GT 2013).

3.4 Heat pumps and CHP

It is not necessary to dig deep to get heat; there is also heat available near the surface. It can be captured and upgraded using electricity-powered heat pumps although, depending on the depth, it will be mainly solar heat, not geothermal, as is obviously the case for air source heat pumps and also for heat pumps using large masses of water as a heat source.

Although there can be problems with condensation and frost, air sourcing is easier (there is no need to bury pipes), but the use of ground-source technology, with pipes installed to capture heat from the ground, is expanding rapidly, with perhaps 200 000 units having been installed in domestic and commercial buildings around the world and many more are planned. The UK for example expects to have over 350 000 in use by 2020.

In theory a heat pump, working like a refrigerator in reverse, can deliver heat with around three times the energy value of the electricity fed in to run it. In practice they may not always achieve these high levels of return, especially air-sourced units in cold, damp weather (Roy *et al* 2010). But heat pumps are useful in off-gas grid locations and do offer a way to upgrade low-grade heat, from whatever source, including solar and geothermal. Moreover, if they are run using electricity from a renewable source, their carbon emissions will be low.

However there is a rival approach to heating. Steam-raising power plants, whatever the heat source/fuel, can be operated in CHP mode, so that some of the heat that would otherwise be wasted in the conversion process is captured for use in DH networks. As I noted earlier, CHP can increase energy conversion efficiency up to 80% or more, compared with around the 35% typical of conventional power plants. There are many CHP/DH projects across the EU and elsewhere (Andrews *et al* 2012, IEA 2008, 2009).

Most current CHP plants use fossil fuel, often gas, but biomass is also being used, with plant load factors of around 90%, as was indicated in table 2.2 in chapter 2. Geothermal power plants can also be run in CHP mode and, as table 2.2 indicated, they can have load factors of 80%. Whatever the heat source, CHP can be attractive if there is a suitable local heat load, e.g. a city or urban area. Domestic scale micro-CHP units are also now available, but larger systems are more efficient: there are economies of scale. That also applies to large-scale heat pumps, which I will be looking at later. It is worth noting, for the sake of comprehensiveness, that there are also gas-fired absorption heat pumps, usually used for large heating projects. Though they are less efficient than the more familiar electric heat pumps, which have electric motor driven compressor systems, gas is cheaper than electricity, and overall fuel savings of 50% have been claimed (ENER-G 2013).

Interestingly, it has been claimed that large community-scaled CHP plants linked to DH networks are far more efficient than domestic-scaled heat electric pumps. CHP/DH systems use heat that would otherwise be wasted so, unlike heat pumps, there is no extra energy input. Heat pumps have a coefficient of performance (COP), relating energy out to energy in, of around 3, but CHP/DH can deliver a COP equivalent of up to 9 or more, depending on the grade of heat that is required (Lowe 2011).

It has been argued that, in primary energy terms, heat pumps would need a COP of 4 (400% efficiency) to compensate for the fact that electricity from the grid was at present produced/delivered so inefficiently (at around 35% fossil/nuclear generation efficiency). Otherwise it would be better to use gas direct in a gas-condensing boiler although, in emission terms, that would not apply if the electricity was from non/low carbon sources.

Running plants in CHP mode, sometimes called 'co-generation', does mean that slightly less electricity is produced since, in practice, to get reasonable temperatures, some steam is taken out of the turbine flow before it reaches the exhaust. But that means that the ratio of heat to power produced can be adjusted, so that CHP plants can be used to balance variable wind generation. They can produce more electricity when wind is low, or produce more heat (possibly for storage) when wind is high (IEA 2011b).

Moreover, if CHP plants use a renewable source of heat, like geothermal or biomass, their carbon emissions, already low per kWh, should be even lower, and cost less per tonne of CO_2 saved than heat pumps, by up to five times or more (Kelly and Pollitt 2009). Though, be warned: there is some debate on this; it may depend on the carbon content of the electricity used by the heat pump and the grade of heat that you want out, although some DH systems can operate at quite low temperatures (Woods 2011, MacKay 2007).

It is also worth noting that, while small heat pumps may be less attractive than CHP/DH, large-scale heat pumps can be better. The UK's Royal Academy of Engineering says that 'the costs of large heat pump installations per kW are a quarter of that for domestic-scale installations' (RAE 2012). There are some large utility-scale heat pumps in use, reportedly achieving high efficiencies. For example, about 60% of the total energy input for Stockholm's heat network is provided by a DH plant with six 1180 MW heat pumps (total heat supply capacity 420 MW(th)) used for base load production along with a (mostly) biofuel-fired backup plant (total heat capacity 200 MW). The heat pumps use the sea as a heat source. Warm surface water is taken during summer, while in winter the water inlet is at 15 m depth where the temperature is at a constant +3°C.

Helsinki in Finland also has a large heat pump plant producing DH with a capacity of 90 MW, as well as cooling, with a capacity of 60 MW, using heat from the sea and from wastewater fed into the sea from a central wastewater treatment plant. CO_2 emissions from the heat pump plant are said to be over 80% less than from using heavy fuel oil or individual cooling compressors in each house (Energy Enviro 2007).

Denmark meanwhile is looking to the use of large heat pumps driven by wind-derived electricity to upgrade heat from CHP plants, and feed it to DH networks. The Danish District Heating Association claims the initiative could make the Danish heating sector CO_2 neutral by 2030 (Building4Change 2012).

Installing DH pipe networks, whether CHP or large heat pump fed, can be disruptive, much more so than installing small heat pumps in individual houses, but once installed and linked to the central heating radiators of houses and other buildings, unlike with domestic heat pumps, there is no in-house device to maintain, and DH pipes can be fed with heat from any source. A DH network is a major, flexible, infrastructure asset, which can be fed with heat from whatever source is currently the most viable.

In a review of heating for building, the UK's Royal Academy of Engineering commented that 'larger district systems, incorporating a CHP facility and providing heating, are significantly more efficient than domestic level installations' (RAE 2012). They also saw advantages in DH, in that central systems 'are likely to offer much greater energy storage than do systems designed for individual households' since 'the mass of water in the underground pipes provides a heat store that evens out daily peaks and troughs in demand. This can be supplemented by hot water tanks to increase energy storage'.

Note also that heat can be sent quite long distances without significant losses. Oslo's district heating network is fed via a 12.3 km pipe from a waste burning plant in the city outskirts. In Denmark there is a 17 km link from a CHP plant to the city of Aarhus. The Helsinki scheme mentioned above is part of a 1150 MW(e) and 3600 MW(th) CHP/DH system, supplying over 93% of Helsinki's heat, including a plant linked in via a 30 km pipe in a tunnel, while in the Czech Republic heat is delivered by a 200 MW capacity heat main to Prague from a power station 65 km away. However, DH does need proper maintenance. Some old Soviet-era systems have not survived well (Loginov 2013).

I will be returning to the issue of heat and electric power infrastructure in chapter 5, which looks at the integration of the various renewable energy options, for heat as well as electricity, into complete sustainable energy systems, with storage being one issue. It also looks at the issue of scale, which I have touched on above. Suffice it to say at this point that, while domestic-scale micro-generation units, including biomass or gas-fired CHP units and heat pumps, are on sale, larger units are usually more efficient. In addition to the basic nature of the devices, this is because of their typical operational patterns. Rather than having to switch on and off to follow the daily energy use patterns in individual houses (something most machines do not like), with larger units supplying heat via DH (if available), demand patterns are averaged out across many houses, so the central plants can run less erratically and more efficiently.

As I have indicated, scale effects are also important in relation to solar heat use and storage. Community-scaled systems, both for supply and storage, have efficiency and economic advantages. As the Royal Academy of Engineering pointed out: 'well insulated hot water tanks or underground inter-seasonal thermal stores will be simpler to provide on a community basis given the small (and reducing) size of most UK homes'.

This chapter has looked at possible renewable options for heat and also electricity supply, based on technologies using heat as their input. Clearly not all of them will be viable everywhere, but some are already delivering large amounts of energy, notably via direct solar heating, at the individual house or community/DH level, while large-scale CSP may well deliver significant amounts of electricity. The IEA says 11.3% of global electricity could be provided by CSP by 2050, and that may be conservative.

Biomass is already a major source, and if the environmental issues can be resolved, it could expand considerably as a source for heat, electricity and also, although possibly with more limits, vehicle fuel. Geothermal could also make a large contribution to heat as well as electricity production. Next, however, I look at the use of light as an energy source for electricity production on possibly a very large scale.

Summary points

- **Biomass** is a well-established heating option, but new crop/biomass waste choices and conversion processes are making its use more efficient and sustainable, including for electricity generation, although land-use and biodiversity issues remain, especially in relation to biofuels for vehicles.

- **Solar heat** is one of the largest and least invasive renewable sources in use globally at present, and solar thermal electricity generation via CSP is becoming established in desert areas. Other solar heat based options for electricity generation are less developed.
- **Geothermal energy** is providing both heat and electricity and is likely to continue to expand around the world, although there are some minor environmental issues.
- **Heat pumps** can play a part and the use of CHP may help get more value from heat, particularly when linked to DH networks on the community scale, offering heat storage and balancing options.

References

Almohsen R 2012 Egypt to test multi-purpose power units, SciDev.Net, May 11, http://www.scidev.net/en/climate-change-and-energy/renewable-energy/news/egypt-to-test-multi-purpose-power-units.html

Alspach K 2013 Joule says it can now produce renewable gasoline & jet fuel without biomass, *Boston Business Journal*, April 15, http://www.bizjournals.com/boston/blog/startups/2013/04/joule-renewable-gasoline-jet-fuel.html

Andrews D, Krook Riekkola A, Tzimas E, Serpa J, Carlsson J, Pardo-Garcia N and Papaioannou I 2012 Background Report on EU-27 District Heating and Cooling Potentials, Barriers, Best Practice and Measures of Promotion, European Commission DG ENER Joint Research Centre, Petten, Report EUR 25289 EN, http://setis.ec.europa.eu/node/2115

BBSRC 2012 'Meet the Gribbles' news report, UK Biotechnology and Biological Sciences Research Council, Swindon, http://www.bbsrc.ac.uk/news/industrial-biotechnology/2012/121128-f-meet-the-gribbles.aspx

BEG 2013 Comments from the Biomass Energy Centre on the 'Dirtier than Coal' report, UK Government Biomass Energy Centre, Farnham, www.biomassenergycentre.org.uk

Beringer T, Lucht W and Schaphoff S 2011 Bioenergy production potential of global biomass plantations under environmental and agricultural constraints, GCB Bioenergy, Potsdam Institute for Climate Impact Research

Biofuelwatch 2012 NGO report on Biomass Energy Carbon Capture (BECCs), Biofuelwatch, http://www.biofuelwatch.org.uk/2012/beccs_report/

Biofuelwatch 2013 NGO evidence to a Select Committee hearing on biomass, Biofuelwatch, http://www.biofuelwatch.org.uk/2013/ecc-biomass-evidence/

Brand B 2010 MENA CSP development overview, Sun and Wind Energy, Dec 5, including interview with Prof. Amin Mobarak, Cairo University

Building4Change 2012 Storing heat energy makes the most of fluctuating wind power, BRE Trust, March 27, http://www.building4change.com/page.jsp?id=1219

Butti K and Perlin J 1980 *The Golden Thread: 2500 Years of Solar Architecture and Technology* (Palo Alto: Cheshire Books)

Chandrasekharam D and Bundschuh J 2008 *Low Enthalpy Geothermal Resources For Power Generation* (London: Taylor and Francis)

CISIRO 2013 Solargas website, http://csirosolarblog.com/tag/solargas/

CNG 2013 CNG Services Ltd, UK promoter of CNG, SNG and green gas grid injection, http://www.cngservices.co.uk/

CSP Today 2013 CSP Today, trade journal/global CSP information website, http://www.csptoday.com/

Desertec 2013 Desertec Foundation website, http://www.desertec.org/

DII 2012 Desert Power 2050: Perspectives on a Sustainable Power System for EUMENA, Desertec Industrial Initiative, http://www.dii-eumena.com/desert-power-2050

Drake 2013 Drake Landing solar storage project, Canada, http://www.dlsc.ca

EERE 2013 Thermal Storage Systems for Concentrating Solar Power, US Department of Energy, Energy Efficiency and Renewable Energy program, http://www.eere.energy.gov/basics/renewable_energy/thermal_storage.html

EGS 2013 Engineered Geothermal Systems company website, http://www.egs-energy.com

Elliott D 2009 The biochar debate *Renew Your Energy* blog, Environmental Research Web, http://environmentalresearchweb.org/blog/2009/10/the-biochar-debate.html

Elliott D 2010 Biochar reviewed *Renew Your Energy* blog, Environmental Research Web, http://environmentalresearchweb.org/blog/2010/09/biochar-reviewed.html

Elliott D 2011 Genetic energy *Renew Your Energy* blog, Environmental Research Web, http://environmentalresearchweb.org/blog/2011/12/genetic-energy.html

Elliott D 2012 Biomass on a slow burn *Renew Your Energy* blog, Environmental Research Web, http://environmentalresearchweb.org/blog/2012/05/biomass--on-a-slow-burn.html

ENER-G 2013 Gas absorption heat pumps, ENER-G Sustainable Technologies, company website, http://www.energ-group.com/heat-pumps/gas-absorption-heat-pumps/

Energy Enviro 2007 Helsinki's heat pump system, Energy Enviro Finland, http://www.energy-enviro.fi/index.php?PAGE=806&PRINT=yes%20More%20www.helsinginenergia.fi

Europa 2013 Energy from abroad: Euro–Mediterranean Partnership (EUROMED), http://ec.europa.eu/energy/international/euromed_en.htm

FAO 2010 Making Integrated Food-Energy Systems work for people and climate, UN Food and Agriculture Organisation, http://www.fao.org/docrep/013/i2044e/i2044e.pdf

Forest Research 2012 Carbon impacts of using biomass in bioenergy and other sectors: forests, Forest Research and North Energy report for the Forestry Commission, www.gov.uk/government/uploads/system/uploads/attachment_data/file/48346/5133-carbon-impacts-of-using-biomanss-and-other-sectors.pdf

Fowles M 2007 Black carbon sequestration as an alternative to bio-energy *Biomass Bioenergy* **31** 426–32

GCCSI 2010 BECCS, Global Carbon Capture and Storage Institute, http://www.globalccsinstitute.com/resources/publications/global-status-beccs-projects-2010

GEL 2013 Geothermal Engineering Ltd company website, http://www.geothermalengineering.co.uk/

GJE 2012 Biofuels: A Burning Problem, GM Biomass issues-NGO Power Point slides, http://www.globaljusticeecology.org/files/biofuels-ppt-web2.pdf

Green Trust 2013 Solar ponds overview, http://www.green-trust.org/solarpond.html

Grydehoj H and Ulbjerg F 2005 District heating—a precondition for efficient use of solar heating, News from DBDH 2, Denmark, www.stateofgreen.com/CMSPages/GetFile.aspx?guid=5c4874f6-dd3c-49fc-b292-f828da607502

GT 2013 GTEnergy geothermal project, company website, http://www.gtenergy.net

Hecht J 2012 Laser drills could relight geothermal energy dreams *New Scientist*, Dec 13

ICAX 2013 ICAX Interseasonal heat stores, company website, http://www.icax.co.uk/interseasonal_heat_transfer.html

IEA 2008 Combined Heat and Power: Evaluating the benefits of greater global investment, International Energy Agency, Paris, http://www.iea.org/publications/freepublications/publication/name,3769,en.html

IEA 2009 Cogeneration and District Energy, International Energy Agency, Paris, http://www.iea.org/publications/freepublications/publication/name,3805,en.html

IEA 2010 Bioenergy, Land Use Change and Climate Change Mitigation, IEA Bioenergy, International Energy Agency, Paris, http://www.ieabioenergy.com/LibItem.aspx?id=6770

IEA 2011a Technology Roadmap: Biofuels for Transport, International Energy Agency, Paris, http://www.iea.org/publications/freepublications/publication/Biofuels_Roadmap.pdf

IEA 2011b Co-Generation and Renewables: Solutions for a Low-Carbon Energy Future, International Energy Agency, Paris, http://www.iea.org/publications/freepublications/publication/name,3980,en.html

IEA 2012 Technology Roadmap: Solar Heating and Cooling, International Energy Agency, http://www.iea.org/publications/freepublications/publication/name,28277,en.html

IGA 2013 International Geothermal Association website and data base, http://www.geothermal-energy.org/geothermal_energy/what_is_geothermal_energy.html

Infinity 2013 Infinity solar/ambient energy fed 'Thermodynamic' Hot water heat pump system, company website, http://www.infinityhomesolution.co.uk/thermodynamics.php

IRENA 2013 Concentrating Solar Power: Technology Brief, International Renewable Energy Agency, Abu Dhabi, http://www.irena.org/menu/index.aspx?mnu=Subcat&PriMenuID=36&CatID=141&SubcatID=283

Jalbuena K 2010 Mars-inspired technology makes PV panels self-cleaning, EcoSeed, http://www.ecoseed.org/technology/13801-mars-inspired-technology-makes-pv-panels-self-cleaning

JRC 2013 Carbon accounting of forest bioenergy, Euopean Commission Joint Research Centre, Institute for Energy and Transport, Ispra, http://iet.jrc.ec.europa.eu/bf-ca/sites/bf-ca/files/files/documents/eur25354en_online-final.pdf

Kelly S and Pollitt M 2009 Making Combined Heat and Power District Heating (CHPDH) networks in the United Kingdom economically viable: a comparative approach, Energy Policy Research Group, Cambridge University: EPRG Working Paper 0925, http://www.eprg.group.cam.ac.uk/wp-content/uploads/2009/11/eprg09251.pdf

Loginov M 2013 Winter in Russia: cold indoors as well as out *Open Democracy*, 16 Jan, http://www.opendemocracy.net/od-russia/mikhail-loginov/winter-in-russia-cold-indoors-as-well-as-out

Lowe R 2011 Combined heat and power considered as a virtual steam cycle heat pump *Energy Policy* **39** (9) 5528–34

Lund J, Freeston D and Boyd T 2011 Direct utilization of geothermal energy 2010 worldwide review *Geothermics* **40** (3) 159–80

MacKay D 2007 *Sustainable Energy without the Hot Air* online book, p147; http://www.withouthotair.com/

Marstal 2013 Marstal solar project website, http://www.sunstore.dk/

Medgrid 2013 Medgrid website, http://www.medgrid-psm.com/en/

Minus7 2013 Minus 7 Solar-fed domestic heat hump system, company website, http://www.minus7.co.uk

MIT 2006 The Future of Geothermal Energy, Massachusetts Institute of Technology, Cambridge, http://mitei.mit.edu/publications/reports-studies/future-geothermal-energy

MSP 2013 Paving the Way for the Mediterranean Solar Plan, EU funded regional project assisting the Mediterranean Partner Countries, http://www.pavingtheway-msp.eu/index.php?option=com_content&task=view&id=34&Itemid=48

NCB 2011 Biofuels: ethical issues, Nuffield Council on Bioethics, London, http://www.nuffield bioethics.org

NRDC 2009 At the Confluence of Water Use and Energy Production, Pierre Bull's National Resources Defense Council Staff Blog, NRDC, Washington, http://switchboard.nrdc.org/blogs/pbull/at_the_confluence_of_water_use_1.html

OTEC 2013 Ocean Thermal Energy Corporation, company website, http://www.otecorporation.com/ocean_thermal_energy_conversion.html

Preheat 2007 Solar Heat Storages in District Heating Networks, Comprehensive list of Danish literature and R&D projects, Ellehauge & Kildemoes, COWI, Preheat study, http://www.preheat.org/fileadmin/preheat/documents/reports/Solar_heat_storages_in_district_heating_networks_ANNEX.pdf

RAE 2012 Heat: degrees of comfort?, Royal Academy of Engineering, London, www.raeng.org.uk/heat

Rein W 2011 Geothermal fracking: what are the risks and potential claims?, ACC/Lexicolgy website, Aug 31, http://www.lexology.com/library/detail.aspx?g=9ab6ec95-683e-49b3-966c-be3dd2140d84

Roy R, Caird S and Potter S 2010 Getting warmer: a field trial of heat pumps, The Energy Saving Trust, London

RSPB *et al* 2012 Dirtier than coal? Why Government plans to subsidise burning trees are bad news for the planet, Friends of the Earth, Greenpeace and the RSPB, http://www.rspb.org.uk/Images/biomass_report_tcm9-326672.pdf

Sahara Forest 2013 Sahara Forest group, Seawater Greenhouses, http://www.saharaforestproject.com

Schlaich J, Bergermann R, Schiel W and Weinrebe G 2005 Design of commercial solar updraft tower systems-utilization of solar induced convective flows for power generation *J. Sol. Energy Eng.* **127** (1) 117–25

Science Daily 2013 Solar booster shot for natural gas power plants *Science Daily*, April 11, http://www.sciencedaily.com/releases/2013/04/130411152332.htm

SDH 2013 EU Solar District Heating Info hub, http://www.solar-district-heating.eu/

Sahakian M 2011 Understanding household energy consumption patterns: When ‘West Is Best’ in Metro Manila *Energy Policy* **39** (2) 596–602

SKM 2012 The Geothermal Energy Potential in Great Britain and Northern Ireland, Report by consultants Sinclair Knight Mertz (SKM) for the Renewable Energy Association (REA), http://www.globalskm.com/Knowledge-and-Insights/News/2012/SKM-report-on-Geothermal-Energy-Potential-in-Great-Britain--Northern-Ireland.aspx

Slade R, Saunders R, Gross R and Bauen A 2011 Energy from biomass: the size of the global resource, UKERC Report, UK Energy Research Centre, London

Solar Done Right 2010 Environmental Impacts of Large-Scale Solar Projects, Green NGO website, http://solardoneright.org/index.php/briefings/post/env_impacts_of_large-scale_solar_projects/

Solar PACES/ESTELA/Greenpeace 2009 Concentrating Solar Power—Global Outlook 09, Greenpeace International/IEA Solar PACES/European Solar Thermal Electricity Association report

Steffersen H 2007 EU aim at great expansion of large-scale solar thermal plants, *DBDH Hot/Cool Journal* **4** 12–15, http://dbdh.dk/images/uploads/pdfbladet/EU%20aim%20at%20great%20expansion%20of%20large-scale%20solar%20thermal%20plants.pdf

Stegen K, Gilmartin P and Carlucci J 2012 Terrorists versus the Sun: Desertec in North Africa as a case study for assessing risks to energy infrastructure *Risk Management* **14** 3–26, http://www.palgrave-journals.com/rm/journal/v14/n1/full/rm201115a.html?utm_source=SilverpopMailing&utm_medium=email&utm_campaign=J%20-%20RM%20DESERTEC%20Article%20%281%29&utm_content=

Sterner M, Pape C, Saint-Drenan Y-M, von Oehsen A, Specht M, Zuberbuhler U and Sturmer B 2010 Towards 100% renewables and beyond power: The possibility of wind to generate renewable fuels and materials, Fraunhofer Institute/IWES, http://www.iwes.fraunhofer.de/de/

publikationen/uebersicht/2010/towards_100_renewablesandbeyondpowerthepossibilityofwindto genera.html
SWG 2013 Seawater Greenhouse website, http://www.seawatergreenhouse.com
Tetzlaff K 2010 Thermo-chemical biomass/wastes processing, Independent Power & Energy Europe/ Claverton Energy Group Conference, Birmingham, June
UKERC 2011 Biomass Resources and Energy Applications, UK Energy Research Centre, London, http://www.ukerc.ac.uk/support/tiki-index.php?page=Biomass+Resources+and+Uses
UKWIN 2013 UK Without Incineration Network, environmental campaign group, http://www.ukwin.org.uk
University of Texas 2012 Analysis of Innovative Feedstock Sources and Production Technologies for Renewable Fuels, Final Report EPA: XA-83379501-0 Chapter 6, http://www.utexas.edu/research/ceer/biofuel/pdf/Report/h_EPA%20Alt%20Fuels%20Final%20Report_Chapter%206_Final.pdf
Veal L 2009 ENERGY: Geothermal Is Not So Clean, Inter-press service News Agency, May 26, http://www.ipsnews.net/news.asp?idnews=46969
Way J 2008 Storing the Sun: molten salt provides highly efficient thermal storage *Renewable Energy World*, June 26, http://www.renewableenergyworld.com/rea/news/article/2008/06/storing-the-sun-molten-salt-provides-highly-efficient-thermal-storage-52873
Wenzel H 2010 Breaking the biomass bottleneck of the fossil free society, CONCITO report, Denmark, Sept, http://www.concito.info/node/317?language=en
WBDG 2012 Passive Solar heating, Whole Building Design Guide, US National Institute of Building Sciences, http://www.wbdg.org/resources/psheating.php
Woods P and Zdaniuk G 2011 CHP and District Heating—how efficient are these technologies?, CIBSE Technical Symposium, DeMontfort University, Leicester, UK, 6/7 September
Worldwatch 2007 Biofuels for transport, Worldwatch Institute, Washington, D.C., Earthscan, London
Zoback M and Gorelick S 2012 Earthquake triggering and large-scale geologic storage of carbon dioxide *PNAS* **109** (26) 10164–8, http://www.pnas.org/content/109/26/10164.short

IOP Publishing

Renewables
A review of sustainable energy supply options
David Elliott

Chapter 4

Light

Energy from light: PV solar and direct conversion

Many of the power and heat technologies I have looked at so far have evolved from versions that were used traditionally, such as windmills, water mills, tidal mills, and biomass combustion. Certainly nearly all use mechanical or thermal energy conversion techniques that were well established, even if they have now been vastly improved or reconfigured, for example to produce electricity. However, there is one technology which is entirely new and based on a conversion process/technique which only emerged as a possibility relatively recently: photovoltaic (PV) solar electricity generation, using sunlight.

It is true that plant photosynthesis converts light energy at very low efficiencies into biomaterials which have an energy content, but photoelectric/photovoltaic devices convert light directly into electricity, with photons releasing electrons in some materials. This is an example of what is called direct conversion. It is not subject to thermodynamic or mechanical limits but, in the case of PV cells and most other forms of electricity-generating direct conversion, the efficiency of conversion is relatively low (10–20%), although much higher than that for photosynthesis. Fuel cells, which I also look at in this chapter, are an exception, they involve the direct conversion of hydrogen to electricity with around 60% efficiency.

Some types of heat- and light-producing devices also involve direct conversion, and these can have very high efficiencies, 80–90%+, resistance heating for example (i.e. electric fires), microwave cookers, and some types of lighting (LEDs). Electric motors are also very efficient energy converters. But in this chapter I am looking at devices which *generate* electricity, not *consume* it, and they are mostly less efficient.

I have focused initially on solar PV since that is the main option so far developed for electricity production from light. Note that is solar *light*, not solar *heat*. Hot sunny weather is not needed to generate electricity, just reasonably strong sunlight, although the stronger it is, the more can be generated. That said, some light is also hot! At the opposite ends of the spectrum of solar radiation that we see as light, we have cold (but potentially biologically damaging) ultraviolet (UV) and hot (and potentially cooking) infrared (IR). So light has a heat component, even if that plays no part in PV conversion,

doi:10.1088/978-0-750-31040-6ch4

except in actually reducing its efficiency. However, it is possible to use the heat part of light, via so-called thermo-electric conversion. But first I focus on PV solar, a key new energy option.

4.1 PV solar

Solar photovoltaics are based on the photovoltaic effect, first discovered in 1839. As noted above, it involves light photons displacing electrons in some materials, so that they create an electric charge and, if connected up appropriately, the cell can produce a current. Interestingly, it was for work on the linked photoelectric (PE) effect that Albert Einstein won his Nobel prize although, as I will explain later, PV differs from PE.

Selenium was used initially, for example in light meters, and later silicon. PV only lifted off as a (small) energy supplier when semiconductor technology developed and PV cells were used to power spacecraft electronic systems. With small volumes of production, they were marginal, and too expensive for large-scale electricity production, until after the 1970s oil crises. Then interest in the technology grew and costs fell (Green 1982).

The attractions of PV are that it is a silent operating, relatively robust and easy-to-fit technology, with no moving parts or plumbing requirements. But since PV cells use specially fabricated materials, they were initially expensive. However, that has changed as volume production has increased and technology has improved. Indeed, PV has one of the best unit cost/installed capacity 'learning curve' slopes in the renewable energy field. Progress down this curve seems likely to continue given that there are many new cell technologies emerging. They will increase energy conversion efficiency and reduce unit cost, since with higher efficiencies less cell material is needed. Depending on the type, commercial silicon cells can have energy conversion efficiencies of 10–18%. More advanced cells, using more exotic materials, can achieve more, at least under laboratory conditions, e.g., the US National Renewable Energy Labs say copper indium gallium selenide (CIGS) solar cells can be almost 20% efficient.

It is of course some way from the lab tests to commercial-scale production, but it may be worth the wait since it is not just efficiency that is important in reducing costs. New cell materials can lead to new, cheaper production methods. Thin film cells, with the cell material deposited in a very thin layer on a substrate backing, are cheaper to make, but tend to have lower efficiencies, so there is a trade-off.

At present, crystalline silicon wafer-based cells still dominate the market, along with thin film cells, but the more advanced CIGS cells are a major new option for larger power uses, although cadmium telluride cell technology is often seen as the most attractive of the new cells. Gallium nitride cells are also seen as very promising (Bhuiyan *et al* 2012). It is a rapidly developing field, with many ideas being developed for so-called third-generation solar cells, including various types of ink dye-based cells, organic/plastic (polymer) cells, and cells exploiting quantum tunnelling effects, all aiming for higher efficiencies and/or lower costs (Boyle 2012).

One of the most advanced solar cells, developed in the USA, is an inverted metamorphic triple-junction gallium indium phosphide/gallium indium arsenide cell which, with light focusing, is claimed has an energy conversion efficiency of around 40% (UDEL 2008). Moving out of the lab, JDSU offer a multiple quantum well enhanced triple-junction GaInP/Ga(In)As/Ge cell, a version of the QuantaSol cell developed at Imperial College London, which is claimed to have 41% efficiency (JDSU 2012).

In advanced designs like this, the light is split optically, with each junction working on different parts of the light spectrum, thus increasing the overall efficiency of energy conversion. They also concentrate the impinging light and are then called concentrating photovoltaic devices (CPVs). In some cases this focusing is achieved by internal lenses, while some devices have larger *external* mirror-type concentrators. Whatever the version, however, the key point is that, while the cell itself may not have 40% or whatever efficiency, the module as a whole can. Focusing allows more solar energy to be collected and focused on a smaller area of the cell. The highest PV efficiency achieved so far is 44% at '947 suns' concentration, with a multi-junction concentrating solar cell (NREL 2012, 2013). Multi-junction cells are preferred since they are less heat sensitive.

Improving cell performance and internal cell optics is not the only way to reduce costs. Other issues include the all-important connectors and packaging/sealing. The quality of the latter can determine how long the unit will continue to work efficiently.

Over the past 30 years, the price per kW of PV modules has reduced by 22% each time the cumulative installed capacity has doubled, and it has been predicted that module costs will fall by about 70% by 2020 and a further almost 80% between 2020 and 2040 (Mott MacDonald 2011). Although there is much optimism, there will no doubt be many problems to resolve before fully commercial third-generation devices emerge. For example, some may need cooling, since performance drops off with temperature. The operational life of some of the new systems is also unknown. Efficiency may fall off with time, as it does with current silicon PV modules. However, Spectrolab's Point Focus metamorphic concentrating cell, which is claimed to offer 40% efficiency, is widely used for satellites, so lifetime survival issues in tough environments have been faced and evidently resolved (Spectrolab 2012).

Focusing systems are only realistic in direct (rather than diffuse) sunlight, so they tend to be used in desert areas. Some systems use flat-sheet Fresnel diffraction lenses to concentrate sunlight, along with tracking mechanisms to orient the cells to follow the Sun across the sky, like Concentrix's 2 MW triple-junction (GaInP/GaInAs/Ge) cell array in Spain. They claim that its energy conversion efficiency is 23% (Concentrix 2013).

In addition to flat-plate tracking systems, large-scale grid-supply CPV systems can also use mirrors or dishes to focus the sunlight, much like the thermal concentrating solar power (CSP) plants I looked at earlier. The basic point is that solar cells are expensive whereas mirrors are cheaper. Until recently, CSP thermal systems seemed to have the cost edge over PV, but now CPV is proving to be more attractive, despite the advantage that CSP has of being able to store heat overnight for 24 hour generation. But then, unlike CSP, CPV does not need cooling water, although in desert temperatures especially that might help, given that cell efficiency falls with increased temperature.

Linked to that last point, some developers have produced hybrid solar thermal/PV systems with, for example, a semi-transparent PV sheet on top of a heat-absorbing solar collector. This PV/T approach can not only keep the PV system cool, but also doubles up on land, roof or wall space usage. One such system, a PV array integrated with a SolarWall air-heating unit, was installed on a roof in the 2008 Beijing Olympic Village. One new CPV system uses extracted heat from cooling for desalination (Nathan 2013).

Several large PV systems are in operation in areas with high insolation. For example there is a 4 MW flat-plate tracking array at Springerville in Arizona, and Abu Dhabi's Masdar 10 MW solar plant is the largest grid-linked PV unit in the Middle East. The largest globally so far is the 600 MW array in northern Gujarat, India, but there are many smaller arrays elsewhere around the world, including many so-called solar farms in the EU, even in the UK (Elliott 2010a). In addition, smaller domestic units have proliferated.

The rapid expansion of domestic-scale and solar farm PV in the EU (by 2012, Germany had 30 GW in place, Italy 15 GW) has not been without problems although, arguably, they were mostly the results of its success. As user uptake rose, due to the feed-in tariffs (FiTs), module prices fell and uptake accelerated further. However, the cost of the FiT subsidies were passed on to all electricity consumers, and it began to rise provocatively, so the tariff levels were cut back; see box 4.1. Nevertheless, deployment has continued on an increasing scale.

Box 4.1. PV solar costs, growth and subsidies

PV solar has been booming in the EU under the impact of premium fixed-price FiT support schemes in Germany, Spain, France, Italy, and to a lesser extent the UK, with volume and markets building rapidly and prices falling significantly, by about 70% since 2008.

The scale of take-up, and the price fall, was so large that from 2011 onwards, FiT levels were cut across the EU more than the planned annual price 'degression' rates, to avoid passing on what some saw as too much extra cost to electricity consumers. Not everyone was happy with the cuts. Some said the tariffs should have stayed high to continue to get prices down and more capacity up, but with the recession, passing on what were seen as excessive FiT costs to consumers was politically contentious. I will be looking at the debate in Germany in chapter 6. There were bitter reactions from the PV and green lobbies to what some saw as excessive cuts, especially in the UK, where a small PV FiT had just been introduced (Elliott 2011, 2012).

Despite this slow-down, the global market for PV is likely to continue to boom. It could reach 130–200 GW by 2015, according to the European Photovoltaic Industry Association. But this will not be just a European expansion. Japan should soon be back in the race; at one time it was a PV leader. Following Fukushima, PV, along with wind, is now a main focus. China has been a major global PV exporter, but it is pushing PV use domestically harder now. It is aiming for 21 GW by 2015. Some analysts even say that, globally, with costs continuing to fall, PV deployment could reach 600–1000 GW by 2020 (McKinsey 2012).

In a report on the prospects for PV in the UK, consultants Ernst and Young suggested that falling solar costs and rising fossil fuel prices could make large-scale PV installations cost-competitive without government support within a decade, and faster given an interim subsidy. Grid price parity with retail prices was expected to be achieved in the UK by 2020 without subsidy for non-domestic installations (Ernst and Young 2011). In other locations it should be earlier. Indeed, PV can be price-competitive with some sources *now*, at some points in the day (Barnham *et al* 2012). For example, at one point in sunny weather in May 2012, PV was supplying around 50% of Germany's electricity, with gas-fired plants having to be shut down to make way.

By the end of 2012 there was 100 GW (peak output) of grid linked PV in place around the world. Initially PV was used for providing electricity directly to buildings but as capacity grew, and small- to medium-sized solar farms on the ground also emerged, it was linked to the grid and, with CPV also developing, grid-linked bulk power supply will expand.

PV can provide electricity for daytime occupancy buildings, but grid linking helps deal with the obvious problem with solar energy: that it is only available in the daytime. In isolated off-grid situations batteries can be used, but this is expensive. Grid linking makes it possible to export and import electricity, on the basis of some type of net trading scheme, sometimes called 'net metering', with FiTs being a variant. Excess output during the daytime, and in the summer especially, is exported to the grid and set against electricity imported at other times, when there is little or no sunlight. The grid thus acts as a store or, more accurately, allows conventional plants to back-up PV, and then be turned down when solar input is available, thus saving fuel and carbon emissions.

However, if new, more economically viable forms of energy storage become available, then the prospects for PV could improve dramatically, both at the small and large scale. For example, if the efficiency of generation of hydrogen gas by electrolysis can be improved, then it may be cheap to store outputs from PV (and from other renewables) as hydrogen, for later use in a fuel cell, to supply electricity when needed; see box 4.2.

Box 4.2. PV storage options

Japanese company Kyocera is selling a system that pairs solar PV panels with lithium-ion batteries for the residential market. The battery storage is rated at 7.1 kWh and weighs about 200 kg. This product emerged, it seems, because of demand for residential backup power supply following the Fukushima nuclear disaster and to take advantage of the new PV FiT in Japan (Solar Selections 2012). But lithium-ion batteries are still expensive, so without extra government incentives, residential energy storage is not likely to take off quickly. Moreover, some would say that having storage at the domestic level is not the best approach. It would be better to feed excess power out on the grid to balance power taken in when there is a shortfall. That, after all, is what FiTs are all about. If storage is needed, it should be done on a larger, more efficient utility scale, via pumped hydro, compressed air, cryogenic air storage, vanadium flow batteries or whatever. I will be looking at these options in chapter 5.

That said, there are seductive small-scale options emerging, like the Fronius energy cell system in which any excess electricity from a PV cell is used to decompose water into oxygen and hydrogen by electrolysis. The hydrogen is then stored, ready to be converted back into electricity in a fuel cell when it is needed (Fronius 2013). I will be looking at fuel cells later in this chapter.

Then again, it might make more sense to set up a full hydrogen economy, with excess electricity from PV and other renewables, large and small, feeding into large, efficient high-temperature electrolysers, and hydrogen then being stored centrally, distributed possibly via the gas grid and used to generate electricity in a large fuel cell, with full heat recovery, via high efficiency CHP generation. I will be discussing the complexities of storage policy in chapter 5.

PV cells are relatively robust but, as mentioned, the lifetime efficiency/long-term performance characteristics of some of the advanced cells are as yet unknown. For example, thin-film cells are potentially prone to degradation. Some cells use toxic materials, so production operations and occupational heath and safety issues have to be carefully monitored. However, once fabricated, environmental impacts are low. Once installed, PV has no emissions, although care must be taken with eventual cell disposal if they contain toxic materials. As with CSP, self-cleaning surfaces are also important to avoid incoming solar energy loss from dust and dirt, and to reduce the need for cleaning with water and detergents (Whitlock 2012, Solar Power Portal 2012).

PV cell manufacture has in the past been relatively energy intensive, so that energy payback times, and thus carbon payback times, have been relatively high compared with some other renewables, but this has been reduced by new cell technology and new high-volume manufacturing techniques. The US National Renewable Energy Labs put energy payback times for multicrystalline modules at up to four years for thin-film modules; paybacks are three years using recent technology (NREL 2004). More recent estimates are even lower, at 1.8 years for monocrystalline and polycrystalline PV, 1.2 years for silicon ribbon and 1.1–0.8 years for cadmium telluride (Fthenakis 2009). Some claim that the new 3D printing techniques may make the fabrication of more efficient cell designs even easier and less energy intensive (Licata 2013).

It seems likely that PV will become increasingly popular as a way to provide electricity for buildings, a roofing and cladding material that actually earns its keep. In addition, PV may begin to play a significant role in bulk electricity supply, not least since PV is well matched to air-conditioning loads, which will grow as the climate changes.

The efficiency of some cells may initially be low, but as the technology continues to develop and markets build, it could be that PV will get so cheap that its cost hardly matters, and given that there are vast areas of desert that could be covered with PV arrays, some optimists are very hopeful for the future. However, there could be major environmental problems with covering large areas of the planet with partially reflective surfaces, quite apart from any impacts on local wildlife. That might also be true of the idea of having floating PV arrays on reservoirs, lakes or even the sea (Levitan 2011), but there is a project underway on a Swiss lake, with the arrays being able to rotate to follow the Sun (Choudhury 2013), and with reservoirs, arrays might help reduce evaporation.

In this context it is interesting that over the years there have been proposals for putting very large PV arrays in deep-space geostationary orbit, perhaps mounted on thin mylar film stretched out over a lightweight frame like a sail, and transmitting the energy to receiving stations on Earth via microwave beams. That would certainly avoid the daytime/night-time problem, and land-use issues, but the costs of this approach (especially launching units into orbit) are likely to be very large, although the US and Japan are both looking at the idea, with some novel designs emerging (Yirka 2012). Given that solar energy falls for some periods on all parts of the planet's surface, it would seem more realistic to try to make use of PV here on Earth, as well as, of course, looking for other terrestrial energy supply options, and in particular other direct conversion technologies for solar light.

4.2 Beyond PV-PE, thermoelectric and photochemical systems

The basis of PV cells is the conversion of the energy of light into electrical energy, i.e. photons into electrons. Plants do something similar, but slowly and inefficiently (they have time on their hands). Human beings evidently are in more of a hurry.

You could try to mimic biological photosynthesis more directly, but improve on it. For example, North Carolina State University researchers have developed a solar cell with a water-based gel infused with light-sensitive molecules, using plant chlorophyll. They say efficiency is low but they felt that 'the concept of biologically inspired "soft" devices for generating electricity may in the future provide an alternative for the present-day solid-state technologies' (PhysNews 2009).

Much further advanced are the various types of dye-based photoelectric (PE) devices. They are sometimes lumped in with PV, but strictly the photoelectric effect is different. It is a surface photochemical effect, rather than occurring in semiconductor materials. The idea is based on the so-called 'artificial leaf' concept originally developed by Swiss Professor Michael Gratzel (Gratzel 2001). He claimed that his photo-electro-chemical cells had an energy conversion efficiency of over 11%, 10 times that of natural systems, and he has continued to develop the idea. Versions have also been developed by, amongst others, G24i in Wales, based on coloured dye, with crystals of titanium oxide, a pigment used in white paint. When hit by sunlight, the dye releases electrons, which are captured by the specks of titanium oxide and can then be used to produce an electric current (G24i 2013). A version using rust, i.e. ferric oxide, is also being developed (Lubic 2013).

One major advantage of ink dye materials is that they can be printed on flexible thin films, producing sheets at up to 100 feet per minute. At present, sensitised ink/dye cells are mainly used for small (e.g. portable) power applications (IDT 2012). However, as I will be explaining below, new applications are opening up. So too are new cell ideas. Some new cells use graphene. This new one-atom-thick carbon-based material looks like having multiple potential applications in the solar cell field, possibly as a replacement for silicon and other semiconductor materials. Researchers at Massachusetts Institute of Technology (MIT) have developed nanowire arrays of graphene by modifying the graphene surface with conductive polymer interlayers (Park *et al* 2012). In parallel, researchers at Stanford University are working on a totally carbon-based thin film cell, with nanotube cathode and graphene anode sandwiching an active layer made of nanotubes and buckyballs, all made by printing or evaporating from inks. When fully developed it is claimed it could provide a tough spray-on PV surface (Ramuz *et al* 2012). There do seem to be many potential uses for graphene; not only does it have very high conductivity and relatively low cost, when it absorbs a single photon it can create multiple electrons, rather than just one (Tielrooij *et al* 2013).

The advent of new ideas like this, and organic/polymer and ink/dye-type cells, opens up new possibilities. Even if their efficiency is low, they are cheap, making it possible to cover large areas, including windows. One early idea was the 'smart window', with a coating which reduces light and heat throughput, and absorbs or reflects some of the energy to power solar cells, which can feed electricity to an air-conditioning unit. An early version was proposed by Professor Keith Barnham at Imperial College London in 2006. Subsequently, a system was developed by Professor Marc Baldo's team at MIT. Large glass panes are treated with a transparent but luminescent material which redirects some of the light energy to solar cells round the edge. So there is a large collector area, but only a few cells (McGrath 2008).

More recently, with cheaper cell material emerging, researchers at MIT have developed transparent organic solar cells, which can be used with windows. They only use near-infrared wavelengths; the rest passes through (Lunt and Bulovic 2011). More recently still, an Oxford University spin-off company has developed a version with the option of different coloured glass (Vaughan 2013), while a Sheffield and Cambridge University team has developed a low-cost 'spray-on' solar coating for windows (Sheffield University 2013). Similarly, a US company has a room-temperature spray-on system using nanoparticles, claimed to cut cost per kW by a third (New Energy Technologies 2013). The efficiencies may be low, but if these 'solar window' systems are as cheap as claimed, there is an obvious potential for wide-scale application, and the advent of these new 'spray-on' materials may reduce solar cell fabrication costs (The Engineer 2012).

In parallel with PV and photoelectric devices, there is a range of thermoelectric devices, which convert heat/thermal radiation into electricity directly. Some thermoelectric materials work by exploiting temperature differences on each side of a semiconductor material. In such devices, electrons move from the hot side to the cold and thus transform heat into electricity. That could be used with solar radiation providing the heat. One problem is that, so far, the energy conversion efficiencies achieved are low. But progress is being made: see box 4.3, which includes technologies using the infrared part of the spectrum. Whether they fit in this chapter or in the previous one is a moot point, since they use solar heat, but they do produce electricity directly, rather than by raising steam.

More exotically still, a University of Utah team has a device that turns solar, or other heat, into sound and then into electricity. Converting heat into sound is the novel first stage. The thermo-acoustic prime mover then drives a conventional piezoelectric device. That produces electricity when squeezed in response to the pressure of the sound waves. The efficiency is low, but some small cooling devices have been built (Utah 2013).

Fascinating though some of these ideas may be (and they do generate electricity without thermodynamic limits, unlike CSP), with these devices we are moving away from light and on to just heat (or even sound). Indeed some of the applications being considered involve the use of waste heat, not solar energy. That may be an important option, but it is moving away from the focus of this chapter. It also raises the issue of whether we should be trying to create electricity in the first place. As seen in the previous chapter, there is a range of options for using heat to produce green fuels, which might be more appropriate.

With that in mind, it is hard to classify (it uses heat from sunlight), but an exotic new solar-based technology may be worth mentioning. This uses solar radiation to produce syngas via ceria redox reactions in a high-temperature solar reactor, with water and carbon dioxide as input feedstock. A solar cavity receiver containing porous ceria felt is directly exposed to highly concentrated solar radiation at a mean solar concentration ratio of 2865 suns. There is then a two-step thermo-chemical cycle. In the first, endothermic (heat in), step at 1800 K, ceria is thermally reduced to an oxygen deficient state. In the second, exothermic (heat out), step at 1100 K, syngas is produced by re-oxidizing

Box 4.3. Thermoelectric and other novel devices

Cells which use the infrared (IR) part of the light spectrum have distinct advantages, since IR is re-emitted by the Earth's surface after sunset, so the device can capture some energy at night and would not be as directionally sensitive as conventional PV cells.

UK company CIP Technologies (CIP), working in partnership with the University of Oxford, say they have achieved 12% energy conversion efficiency for what they call thermo-photovoltaic (TPV) cells. They use first-generation single-junction cells based on indium phosphide materials, which absorb IR heat radiation. They are looking at second-generation cells with a more complex, multi-layer construction to improve IR capture further. This could raise energy conversion efficiency to over 15% (CIP 2013). That would put them on a par with Si PV.

Nanoparticle absorbers open up potentially remarkable new possibilities. Researchers at MIT have developed a flat-panel 'enhanced nanostructured' solar thermo-electric device that can heat water, while Rice University has developed a solar thermal system using nanoparticle absorbers which heat up in sunlight and flash off steam when immersed in water. They claim that the system could have overall energy conversion efficiency of 24% (Rice 2012).

The US Department of Energy's Idaho National Lab has developed a cell that works on IR radiation. Tiny nanoscale antennas resonate when hit by light waves, generating an alternating current at very high frequency, which has to be rectified to be useful (Graham-Rowe 2010). That is tricky at the nanoscale, and there is some way to go before the high efficiencies claimed (60–70%) as ultimately possible can be attained, but researchers at the University of Connecticut and Penn State Altoona seem to be making progress (Fong 2013).

Conventional silicon PV cells can also be made to work in the IR part of the spectrum by interstitial lattice modification with sulphur molecules, producing what is called 'black silicon', with an IR-absorbing surface layer. Lab efficiencies of 18.2% have been reported (Yuan and Branz 2012), but so far they are mainly used for IR detection/measurement, although given the potentially high productivity, commercial power cells may emerge (Davies 2013).

ceria with a gas mixture of H_2O and CO_2. By contrast to direct thermolysis (e.g. high-temperature thermochemical processing), it is claimed that these cycles bypass the CO/O_2 and H_2/O_2 separation problem and operate at lower temperatures, although as yet the process is not very efficient (Furler 2012).

Also hard to classify are various photochemistry ideas, some aiming at hydrogen production (e.g. photo-electrolysis). For example, researchers at the University of Rochester, USA, are trying to develop solar-driven systems for generating hydrogen using complex light-sensitive natural molecules called chromophores and membranes infused with carbon nanotubes/graphene to ensure that the freed electrons are not reabsorbed by the chromophore (Long *et al* 2012). Artificial photosynthesis, using a range of techniques, seems to be an exciting new option (Styring 2012, Faunce *et al* 2013).

There may be prospects for some novel thermochemical/photothermal devices using solar energy, but in terms of my focus in this chapter, are there any *other* clever new direct conversion technologies which might enable us to convert light into electricity efficiently, and on a large scale? Maybe: photochemistry may yet shade into photobiology, and there may be even more exotic options possible for using solar radiation. For example, there are energy components in the wider ranges of the electromagnetic radiation spectrum, beyond visible light and infrared, in radio waves, microwaves, x-rays and gamma rays, whether naturally produced or man-made. We already use some of these forms of radiation to transmit information (wireless radio) or even energy (in microwave beams). In theory we might also be able to convert some of this energy into other forms, although costs may be high and efficiency low.

4.2.1 Direct conversion: beyond photons

This is still all about photons, in their various radiation guises, which would presumably be used either to create heat or to create electrical charge. Our primary source remains the Sun. But it may be worth thinking speculatively for a moment about what that means. It is all about energy conversion. Ultimately what humankind wants, in terms of energy, is light, heat, electricity (for electronics) and motive power (for motors and vehicles). As I have recounted, the story so far is that we have used stored solar energy in fossil material to make electricity by burning it, or just used it for heat. We have used fossil heat or electricity to drive vehicles, and we have used electricity to make light (in light bulbs, LEDs, plasma discharges, fluorescent tubes and so on).

In this book I have looked at new ways of doing all of those things, or at least at new energy sources for powering those activities and processes, based mostly (tidal and geothermal apart) on direct or indirect solar heat (solar thermal, wind, wave, hydro) and light (PV, PE etc). However, until recently these ideas have mostly been ignored or downplayed. Instead, realising that fossil fuels were limited and messy, humankind, being perverse, has tried to find other stored sources of energy, and has turned to uranium, very cleverly releasing some of the vast energy from within the atomic nucleus by fission. A lot of that energy comes out as light or at least gamma radiation, but we use it as heat and then convert that, laboriously, to electricity, much as we did with fossil fuels. However, arguably, we end up with an even more unpleasant mess in the form of radioactive waste.

If controlled nuclear fusion ever becomes possible on a significant scale, we would probably do the same, just use it for raising steam for electricity generation. Most of the energy from high-temperature fusion reactions would come out as fast-moving neutrons, plus some light. We would convert that into heat and then boil water. So it would be an expensive and somewhat odd way to boil water. There might be less mess, though the neutron flux would activate the metal surfaces and internal equipment, so that would have to be stripped out periodically, since otherwise the hard gamma/x-rays produced might interact with the fusion process. So it is not free from radioactive waste. And the next stage is rather old-fashioned: the energy in the neutron flux would be absorbed by a surrounding blanket of lithium, and the resultant heat would be extracted by a network of water pipes running through the blanket (Elliott 2009a, 2010b).

Can we do better? The Sun uses fusion to create heat and light in vast amounts. We can certainly use that. It is a gigantic fusion reactor in the sky. It is free and safely a long way off, and does not need our help with fuel, maintenance, security or waste management. However are the only options to simply use its light and heat, that is, photons?

There are some energy conversion technologies that are based on protons, positively charged sub-nuclear particles which, in their simplest form, are hydrogen nuclei (one proton). Nature uses them in some biochemical processes and there have been attempts to modify them to produce electricity, as in plant-microbial fuel cells (Harris 2013). That is some way off on any scale, but conventional chemical fuel cells are well developed and widely deployed. Some use proton exchange membrane technology to produce electricity, which is just another way of moving electric charge around, with particles that are much larger than electrons.

Fuel cells produce electricity from hydrogen fuel at around 60% efficiency; the rest is converted to heat. If some of that heat can be captured (so it would be fuel cell combined heat and power (CHP)), overall efficiency can be raised, and with improved technology, it can perhaps reach up to 85%. Although most run on hydrogen, some units have integral chemical reformers, so that they can run on methane, while some can run directly on methane with, interestingly, graphene looking like a good new option for this (Paneri *et al* 2012).

The first fuel cell was demonstrated by Sir William Grove in 1839 and they were then developed by Francis Bacon in the 1950s. Later they were used in spacecraft, e.g. Apollo. Now they are a useful tool for our new energy system, making it possible to generate electricity using fossil or green/biogas, or stored wind-derived hydrogen, all without combustion. They are already widely used at various scales, from running laptops to utility-scale electricity generation. For example, there is a 14.9 MW fuel cell power plant in Bridgeport, Connecticut. Cells for vehicle use are a big new market, but in the stationary market, it could be that electricity (and heat) generation at the domestic scale (micro-CHP), will lift off. Some larger-scale fuel cell projects have also been developed, running on hydrogen produced by electrolysis using wind/PV-derived electricity (HARI 2006, Hydrogen Office 2013, Pure Energy 2013). That idea is now being expanded in some of the 'power to gas' projects in Germany and elsewhere, which I will be looking at in the next chapter. As I mentioned earlier, it is conceivable that hydrogen, for use in fuel cells, could be produced by direct high-temperature solar thermal dissociation of water (using highly focused solar), although that is inefficient and more likely some catalytic process would be required (DoE 2012). But high-temperature plasma-based systems have been devised, working on water vapour (Hion 2013), though that is still a long shot.

Fuel cells offer us one way to use protons (from hydrogen), but there is also another direct energy conversion option which might provide a way to use protons: magneto-hydrodynamic (MHD) technology. Magnetic fields can induce currents in a moving conductive fluid, which in turn create forces on the fluid and also change the magnetic field itself. In theory MHD could be used to convert the energy in a fast-moving stream of charged particles (e.g. protons) into electric current. At present, however, there is no obvious source of fast-moving charged particles (at least on Earth), unless we use a lot of energy to create them, but it is conceivable that the neutron stream from a fusion reactor could be converted into protons and accelerated through an MHD system to produce electricity.

However, fusion is a long way off, at best many decades. So this route offers no help at all for dealing with our pressing climate and energy security concerns, whereas we have renewables now, including PV and other photoconversion options. So perhaps it does not matter that, fuel cells apart, the use of protons is limited. We do have the Sun and its photons. Direct conversion of this large energy source via PV, PE and so on can supply electricity in large quantities around the world.

There are limits, but even in the often cloudy UK, the Department of Energy and Climate Change's DECCs Pathways analysis has suggested that a PV capacity of up to 22 GW might be possible by 2020. Moreover, based on German experience, it has been claimed that, given proper support, UK PV could expand to 37 GW by 2020 (Barnham *et al* 2012). In more favourable climates, very much more is possible. Shell's 2013 'Oceans' scenario has solar as the largest single energy contributor globally by 2060 (Shell 2013). We could be moving to a solar world.

4.3 One world under the Sun: a global grid?

To reprise the story so far, the prospects for solar energy are very good. Globally, the planet receives 90 000 TW of solar energy annually. Nature converts some of this to biomass via photosynthesis in a slow, inefficient process. We can and do use some of the results for energy production, as well of course as for food. We have also used the stored solar energy that had been laid down over millennia in the form of fossil carbon material, in recent times, burning off 1 million years' worth of this stored solar energy each year.

Fortunately, we are learning to do better, by converting solar heat and indirect solar-derived hydro, wind and wave energy into electricity. And as this chapter shows, we have also learnt how to convert light directly into electricity, with a large energy potential.

It does of course have limits. Manufacturing solar cells is energy intensive. However, as I have indicated, the energy intensities are falling. In terms of 'energy debts', one way to compare embedded energy intensities is to look at 'energy return on energy invested' (EROEI), i.e. the total lifetime energy output divided by the energy input needed for manufacture and running, the latter being zero for renewables like PV. A study by Harvey put EROEIs for PV at 8–25, depending on type and location. So you can get up to 25 times more energy out than you have to use to make the cells. For comparison, he puts EROEIs for CSP at 8–40, depending on type (parabolic troughs were seen as the best) and wind at 40–80 (Harvey 2010). These EROEIs should all improve with new technology. For example in an earlier assessment using older data, Gagnon (2008) quotes EROEIs for PV only in the range 3–6 and for wind at ~18 for offshore and 34 for on land.

EROEIs depend on the operational lifetime of the plant. The estimates above typically assume 20 years. However, once built, hydro dams last for centuries and hydro plants only need occasional turbine refurbishment, so it is not surprising that for hydro Gagnon quotes EROEIs of 200 or more. The embedded energy offers very good returns.

By comparison, despite their relatively long operating lifetimes (40 years perhaps), nuclear plants have quite low EROEIs, in part since energy is needed to extract and process the uranium fuel. Gagnon puts the EROEI for current pressurised water reactors at 14–16. However, this will fall as and when lower-grade ores have to be used. Harvey quotes 16–18 for plants using the current world average uranium ore grade of 0.2–0.3%, but sees this as falling, for an ore grade of 0.01%, to 5.6 for underground mining and to 3.2% for open pit mining, and to as low as 2 for *in situ* leaching techniques. That would put nuclear well below the EROEIs Harvey quotes for coal (5–6.7) and gas plants (2.2).

Gagnon puts the EROEI for conventional coal-fired plants with carbon capture and storage (CCS) even lower, at only 1.6, if the coal is transported 2000 km. Clearly the renewables, including PV, can do much better than that.

PV may not be the best renewable in current EROEI terms, and large areas have to be used to extract significant amounts of energy, but as well as rooftops, there are very large desert areas that can be used. I have already noted the significant land-use implications of biomass use, and I will be looking at land-use issues more generally in chapter 6, but it is worth pointing out here that, on some estimates, while biomass is very land hungry (around 533 km^2 TWh^{-1}) and hydro quite high (152 km^2 TWh^{-1}), PV's land-use is 45 km^2 TWh^{-1}, while that for wind is 72 km^2 TWh^{-1} (Gagnon *et al* 2002). The wind/PV comparison is, however, difficult, since much of the PV could be on rooftops, while most of the land around wind turbines can be used for other purposes. Moreover, while land for wind may be more constrained than land/rooftops for PV, the offshore wind resource is very large. Even so, if desert areas are used, then the solar resource is very large.

4.3.1 The global spread of PV

So far, in part due to its initial high energy intensity, PV has been expensive, but with new technologies emerging and markets building, costs and energy intensities are falling. One study even suggested that PV had reached the point when it was competitive with new nuclear, at least in parts of the USA, with subsidies factored in (Blackburn 2010). CPV is also growing, reaching around 1 GW by 2012 (CPVC 2013).

As I noted earlier, the rapid take-up of PV by domestic consumers has been stimulated in much of the EU by the availability of FiTs, offering guaranteed prices for electricity produced by consumers who had installed PV. This had led to a new situation in some countries where the large power companies/utilities are no longer completely in control of the market or indeed of the power system.

Consumer-led PV uptake thus represents an archetypal 'destructive innovation', challenging and changing market and industrial paradigms. Schleicher-Tappeser, writing in *Energy Policy*, says that it allows consumers of all sizes to produce power themselves: 'new actors in the power market can begin operating with a new bottom-up control logic'. He calls them 'prosumers'. He adds that the 'increasing autonomy and flexibility of consumers challenges the top-down control logic of traditional power supply and pushes for a more decentralised and multi-layered system' (Schleicher-Tappeser 2012).

It will be interesting to see if it plays out the same way elsewhere. For example, following the so-called 'Arab Spring', some of the new governments in North Africa and the Middle East are pushing ahead with solar energy, both CSP (see box 3.3, chapter 3) and PV, although grass-roots pressures do not seem to be the leading factor. Instead a new reality in oil economics seems to be a strong influence, at least in some locations. For example, with domestic energy use rising, many Gulf states are looking at solar energy as a way to diversify from oil for commercial reasons (so as to release oil for sale), quite apart from anything directly to do with climate change or social policy. It is only on a relatively small scale so far, but in addition to the CSP projects in the region, the UAE has a plan for a 100 MW PV array, and many more are expected. Dubai is planning a 1 GW project and Saudi Arabia an even larger one.

Certainly, there is an incentive to go for PV as an alternative to using oil to generate electricity, demand for which is rising rapidly as affluence increases, for example due to wider use of air-conditioning, and also the growth of desalination of water. That is beginning to eat into potential oil export revenues. One study suggested that, 'on the current trajectory, Saudi Arabia's domestic energy consumption could limit its exports of oil within a decade' (Lahn and Stevens 2011). Moreover, longer term, there is the potential for creating a new export industry as oil reserves dwindle. Climate change is also a real issue in these already overheated areas. It will push up energy use for cooling even more. So a new approach to energy may well emerge.

I will be looking at the take-up of PV and other renewables around the world in chapter 6. But clearly, if prices can be brought down, PV will be eminently suited to many developing countries. More generally, large-scale direct conversion of light into electricity, if cheap and easy, could be a 'game changer' globally.

An intriguing idea from researchers at the University of Tokyo is to establish a self-sustaining Sahara Solar Breeder operation, using desert sand to produce PV units and desert sun to generate electricity to power production. Energy generated by the first wave of plants would be used to 'breed' more silicon manufacturing and solar energy plants. They in turn would be used to breed yet more in a 'self-replicating' system. Certainly, the MENA region is not short of sand, space or sunlight.

Launched in 2010, a five-year $1.2 million research initiative was funded by the Japanese government under the auspices of the International Research Project on Global Issues, with the aim of exploring the possibility of manufacturing high-quality silicon from desert sand and of building a high-temperature superconducting long-distance DC power supply system. The researchers suggest that, if all goes well, this approach could, in theory, make it possible to generate 50% of global electricity from solar by 2050, distributing solar energy globally through a superconducting supergrid (DigInfo 2010).

Clearly this is all very speculative, but it is interesting, the last point especially. Solar energy has the major drawback that it gets dark at night, at least for half the planet. I mentioned some energy storage options earlier and, while they might help, the perhaps utopian solution as proposed by the Japanese team is a global supergrid. That, crucially, would be able to transmit electricity from the sunny side of the planet to the dark side, an idea that certainly catches the imagination—a global grid enabling PV solar to be used everywhere 24/7. It could also conceivably offer a transmission route for balancing the output of other variable renewables and varying demand around the world (GENI 2013).

4.3.2 A global grid—solar on tap

The global grid may not be so fanciful as it first seems. There are already plans for long-distance high voltage direct current (HVDC) supergrids to link up countries and their renewables, in part to help compensate for the local variability of the resource. I will be looking at that more in the next chapter. Why not go the whole hog and, following in the footsteps of the visionary US thinker Buckminster Fuller, link up continents and the entire planet with a global supergrid? The Japanese researchers mentioned above talked of using superconducting supergrids. That technology may be some way off. But standard

HVDC transmission already exists; it is widely used around the world over long distances and it could be suitable for much longer distances (Chatzivasileiadis *et al* 2013).

A global grid would involve large new grid networks, which could be invasive and also vulnerable. So fans of Nikola Tesla, the visionary US-based electrical engineer (who invented the now ubiquitous alternating current system), might say: why not try his idea of, in effect, establishing a global standing wave of electromagnetic radiation in the troposphere to carry power, with the Earth itself as the return loop? (Tesla 1900, 1914).

Although there are some very short-range magnetic resonance devices for power linking, Tesla's ideas for a global 'wireless electrical power transmission grid' sound like science fiction, with all sorts of possible dangers. They would be even worse (if we are looking at mega, global schemes) than some of the wilder proposals for planet-wide geo-engineering put forward as 'emergency' responses to climate change. For example, it has been suggested that sunlight should be blocked with deep-space orbital screens or artificial reflecting microparticles in near-Earth orbit, or by spraying aerosols into the atmosphere (Royal Society 2009). Painting rooftops white to reflect more sunlight is one of the less risky options, and chemical 'air capture' of CO_2 does have the advantage that, unlike CCS (where capture has to be done at or near power plants), it can be done anywhere (IMechE 2009).

I do not want to get sidetracked into a discussion of these ideas and their problems, but it is worth noting that, for air capture, the concentration of CO_2 in the air will be much less than in power station exhausts, and many of the global geo-engineering schemes have the potential to impose major environmental impacts, some of which might be irreversible.

Moreover the idea, in some of them, of trying to block sunlight should give us a hint, firstly, of how badly we have treated this planet, by burning off about half of the accessible stored solar energy in fossil fuels so far, and secondly, of what a better solution would be, that is to use the sunlight we receive in real time, not block it! Tragically, if instead geo-engineering is backed on a significant scale, that might divert attention and resources from renewable energy development. Although some apologists argue that geo-engineering may be necessary, since renewables will not be developed in time, this could become a self-fulfilling prophecy (Elliott 2009b).

While the sunlight-denying geo-engineering proposals, and also Tesla's ideas, may seem too wild, by comparison the global grid may seem somewhat less far-fetched. We do have a global internet, so there is some sort of precedent. There have been studies of possible intercontinental HVDC electricity grid links, e.g. from the EU via Greenland to North America (Claverton Energy Group 2012). Renewables like PV will work fine without it, but global grid enthusiasts say that, with it, they might work even better.

For the moment that is some way off, although the beginnings of a global link-up are being made, with national grids across the EU and even more widely, being synchronised and integrated. The existence of large untapped renewable resources in remote areas may drive this process further, as for example with the Gobitec CSP supergrid project in Asia, and the Desertec CSP/supergrid initiative in North Africa (DII 2013). These involve solar resources in desert areas and/or wind resources in Morocco and Mauritania (Czich 2011). But there are also proposals for building links from the EU to Russia's huge wind resources in Siberia (Boute and Willems 2012).

HVDC grids, linking to existing grid networks, are not as fanciful as, say, microwave links up, between and then down from a network of geostationary satellites around the world, so we may yet see a global grid emerging incrementally, in which case the world could look very different in the future.

However, leaving very speculative ideas like this aside, I will now move on in the next chapter to look at some more practical and immediately realisable ideas about how to link renewables up in an integrated and balanced energy system.

Summary points

- **PV solar** is gaining ground rapidly, with prices falling, and it does not take much space (on rooftops), though solar farms are more invasive, as are large CPV arrays in deserts.
- **Thermochemical** and **thermo-photovoltaic** cells may offer new options for using solar light and heat.
- **Fuel cells** offer an energy conversion option which can add flexibility to energy systems, and we may yet come up with novel ways to mimic and upgrade photosynthesis.
- There may be entirely new options for using or transmitting **electromagnetic radiation**, but we should not get too carried away with geo-engineering schemes that could have large global impacts, though the HVDC **global power supergrid** idea looks fascinating!

References

Barnham K, Knorr K and Mazzer M 2012 Benefits of photovoltaic power in supplying national electricity demand *Energy Policy* **54** 385–90

Blackburn J 2010 Solar and nuclear costs the historic crossover, NC Warn website, July, http://www.ncwarn.org/?p=2290

Bhuiyan A, Sugita K, Hashimoto A and Yamamoto A 2012 InGaN solar cells: present state of the art and important challenges *IEEE Photovoltaics* **2** 6204302

Boute A and Willems P 2012 RUSTEC: Greening Europe's energy supply by developing Russia's renewable energy potential *Energy Policy* **51** 618–29

Boyle G 2012 Solar photovoltaics *Renewable Energy*, ed G Boyle (Oxford: Oxford University Press)

Chatzivasileiadis S, Ernst D and Andersson G 2013 The global grid *Renewable Energy* **57** 372–83

Choudhury N 2013 CSP floating labs under construction in Switzerland, Jan 29, http://www.pv-tech.org/news/csp_floating_labs_under_construction_in_switzerland

CIP 2013 CIP Photonics, TPV data sheet, http://www.ciphotonics.com

Claverton Energy Group 2012 Prospects for trans-atlantic undersea power transmission, Claverton Energy Group, http://www.claverton-energy.com/prospects-for-trans-atlantic-undersea-power-transmission.html

Concentrix 2013 Concentrix Concentrating solar technology: company website, http://www.concentrix.com/

CPVC 2013 Concentrated Photovoltaics Consortium, www.cpvconsortium.org/

Czich G 2011 *Scenarios for a Future Electricity Supply* (London: IET)

Davies S 2013 If the price is right *Engineering and Technology Magazine* **8** (2) Feb 11, http://eandt.theiet.org/magazine/2013/02/making-the-most-of-it.cfm

DigInfo 2010 DigIngfo TV coverage of Sahara Solar Breeder project, http://www.diginfo.tv/2010/11/24/10-0135-r-en.php

DII 2013 Desertec Industrial Initiative website, http://www.desertec.org/

DoE 2012 High-Temperature Water Splitting, US Department of Energy, http://www1.eere.energy.gov/hydrogenandfuelcells/production/water_splitting.html

Elliott D 2009b Geoengineering—a last ditch response to climate change? *Renew Your Energy* blog, Environmental Research Web, Sept 19, http://environmentalresearchweb.org/blog/2009/09/geoengineering-a-last-ditch-r.html

Elliott D 2009a Why fusion? *Renew Your Energy* blog, Environmental Research Web, July 4, http://environmentalresearchweb.org/blog/2009/07/why-fusion.html

Elliott D 2010a Solar Farming *Renew Your Energy* blog, Environmental Research Web, Sept 30, http://environmentalresearchweb.org/blog/2010/09/solar-farming.html

Elliott D 2010b Fusion - funding problems and choices *Renew Your Energy* blog, Aug 7, Environmental Research Web, http://environmentalresearchweb.org/blog/2010/08/fusion-funding-problems-and-c.html

Elliott D 2011 Fighting for FiTs *Renew Your Energy* blog, Environmental Research Web, Feb 12, http://environmentalresearchweb.org/blog/2011/02/fighting-for-fits.html

Elliott D 2012 PV Solar battle continues *Renew Your Energy* blog, Environmental Research Web, Jan 7, http://environmentalresearchweb.org/blog/2012/01/pv-solar-battle-continues.html

Ernst & Young 2011 UK solar PV industry outlook, Ernst & Young Consultants report for the UK Solar Trade Association, http://www.oursolarfuture.org.uk/wp-content/uploads/The-UK-50kW-to-5-MW-solar-PV-market-190611-Final.pdf

Faunce T, Lubitz W, Rutherford W, MacFarlane D, Moore G, Yang P, Nocera D, Moore T, Gregory D, Fukuzumi S, Yoon K, Armstrong F, Wasielewski M and Styring S 2013 Energy and environment policy case for a global project on artificial photosynthesis *Energy Environ. Sci.* **61** 695–8

Fong P 2013 New Nanoscale Antennas Could Boost Solar Energy Harvesting Efficiency to 70% *Txchnologist*, Feb 22, http://txchnologist.com/post/43730365283/new-nanoscale-antennas-could-boost-solar-energy?utm_source=Txch%2Bnewsletter&utm_medium=email&utm_campaign=email%2Bdistro

Fronius 2013 Fronius Energy Cell: company website, http://www.fronius.com/cps/rde/xchg/SID-316A5419-05794498/fronius_international/hs.xsl/83_19199_ENG_HTML.htm

Fthenakis V, Kim H, Raugel M and Kroner J 2009 Update of PV energy payback times and life-cycle greenhouse gas emissions, 24th European Photovoltaic Solar Energy Conference and Exhibition, Hamburg, Sept 21–25.

Furler P, Scheffea J and Steinfeld A 2012 Syngas production by simultaneous splitting of H_2O and CO_2 via ceria redox reactions in a high-temperature solar reactor *Energy Environ. Sci.* **5** 6098

G24i 2013 G24i company website, http://www.g24i.com/

Gagnon L 2008 Civilisation and energy payback *Energy Policy* **36** (9) 3317–22

Gagnon L, Bélanger C and Uchiyama Y 2002 Life cycle assessment of electricity generation options; the status of research in year 2001 *Energy Policy* **30** (14) 1267–78

GENI 2013 Global Energy Network Institute, San Diego, http://www.geni.org/

Gobitec 2013 Gobitec project website, http://www.gobitec.org/

Graham-Rowe D 2010 Is night falling on classic solar panels? *New Scientist* **20** December

Gratzel M 2001 Photo-electro-chemical cells *Nature* **414** 338–44

Green M 1982 *Solar Cells* (New York: Prentice-Hall)

Green Trust 2013 Solar Pond Briefing information, http://www.green-trust.org/solarpond.htm

HARI 2006 Hydrogen and Renewables Integration project, Prof Tony Marmont, West Beacon Farm/Centre for Renewable Energy Systems Technology, Loughborough University, UK, http://www.ieahia.org/pdfs/HARI.pdf

Harris S 2013 Green electricity roof could provide power, water storage and insulation, Environmental Research Web, March 7, http://environmentalresearchweb.org/cws/article/news/52643

Harvey D 2010 *Carbon-Free Energy Supply* (London: Earthscan)

Hion 2013 Hion solar website, http://www.hionsolar.com/n-hion96.htm

Hydrogen Office 2013 Wind-Hydrogen project, Fife, Scotland, http://www.hydrogenoffice.com/

IDT 2012 Dye sensitized solar cells: technologies markets and layers, IDT report, http://www.idtechex.com/research/reports/dye-sensitized-solar-cells-technologies-markets-and-players-2012-2023-000328.asp

IMechE 2009 Geoengineering, Policy statement Institution of Mechanical Engineers, London, http://www.imeche.org/knowledge/themes/environment/climate-change/geo-engineering

JDSU 2012 JDS Uniphase Corporation website, http://www.jdsu.com

Lahn G and Stevens P 2011 Burning Oil to Keep Cool: The Hidden Energy Crisis in Saudi Arabia, RIIA, London, Chatham House Programme Report, December, http://www.chathamhouse.org/publications/papers/view/180825

Levitan D 2011 More Offshore Ideas: Floating Solar Panels *IEEE Spectrum*, Feb 28, http://spectrum.ieee.org/energywise/green-tech/solar/more-offshore-ideas-floating-solar-panels

Licata J 2013 How 3D printing could revolutionise the solar energy industry *Guardian*, Environment Blog, Feb 22, http://www.guardian.co.uk/environment/blog/2013/feb/22/3d-printing-solar-energy-industry?CMP=twt_fd

Long R, English N and Prezhdo O 2012 Photo-induced charge separation across the graphene-TiO2 interface is faster than energy losses: a time-domain ab initio analysis *J. Am. Chem. Soc.* **134** 14238–48

Lubic N 2013 Midnight sun: How to get 24 hour solar power *New Scientist*, 2901, Jan 23, http://www.newscientist.com/article/mg21729011.900-midnight-sun-how-to-get-24-hour-solar-power.html?full=true

Lunt R and Bulovic V 2011 Transparent, near-infrared organic photovoltaic solar cells for window and energy-scavenging applications *Appl. Phys. Lett.* **98** 113305

McGrath M 2008 Solar dyes give a guiding light, BBC website, July 11, http://news.bbc.co.uk/1/hi/technology/7501476.stm

McKinsey 2012 Solar Power: Darkest before the Dawn, McKinsey consultants, http://www.mckinsey.com/Client_Service/Sustainability/Latest_thinking/Solar_powers_next_shining

Mott MacDonald 2011 Costs of low-carbon technologies, report for the Committee on Climate Change, May, http://hmccc.s3.amazonaws.com/Renewables%20Review/MML%20final%20report%20for%20CCC%209%20may%202011.pdf

Nathan 2013 High-Concentration Photovoltaic Thermal System From IBM Promises 80% Efficiency, Potable Water, And Air Conditioning, Cleantechnia, April 24, mhttp://cleantechnica.com/2013/04/24/high-concentration-photovoltaic-thermal-system-from-ibm-promises-80-efficiency-potable-water-and-air-conditioning/#J8BOH2XVmb0xyiTB.99

New Energy Technologies 2013 US New Energy Technologies Inc., website, http://www.newenergytechnologiesinc.com

NREL 2004 What is the energy payback for PV?, PV FAQS Sheet, US National Renewable Energy Labs, Golden, Dec

NREL 2012 Award-Winning PV Cell Pushes Efficiency Higher, US National Renewable Energy Labs, Golden, Dec 28, http://www.nrel.gov/news/features/feature_detail.cfm/feature_id=2055

NREL 2013 Solar cell efficiency comparison chart, US National Renewable Energy Labs, Golden, http://www.nrel.gov/ncpv/images/efficiency_chart.jpg

Paneri A, Ehlert G, Sodano H and Moghaddam S 2012 Graphene-based Proton Exchange Membrane (PEM) for Direct Methanol Fuel Cell (DMFC), Tech Connect World, http://www.techconnect-world.com/Nanotech2012/a.html?i=1399

Park H, Chang S, Jean J, Cheng J, Araujo P, Wang M, Bawendi M, Dresselhaus M, Bulović V, Kong J and Gradečak S 2012 Graphene cathode-based ZnO nanowire hybrid solar cells *Nano. Lett.* **13** 233–9

PhysNews 2009 Leaf-like solar cells: water-based 'artificial leaf' produces electricity *Physics News*, http://www.physnews.com/nano-materials-news/cluster102685380/

Pure Energy 2013 UK hydrogen/wind/pv/fuel cell system development centre, http://www.pure-energycentre.com

Ramuz M, Vosgueritchian M, Wei P, Gao Y, Yingpeng Wu, Chen Y and Bao Z 2012 Evaluation of solution-processable carbon-based electrodes for all-carbon solar cells *ACS Nano* **6** (11) 10384–95

Rice 2012 Solar steam device, Rice University Nano-Photonics Lab, http://news.rice.edu/2012/11/19/rice-unveils-super-efficient-solar-energy-technology/

Royal Society 2009 Geoengineering the climate The Royal Society, London, http://royalsociety.org/policy/publications/2009/geoengineering-climate/

Schleicher-Tappeser R 2012 How renewables will change electricity markets in the next five years *Energy Policy* **48** 64–75 September

Sheffield University 2013 Solar cell fabrication is simplified by spray painting, Sheffield University press release, Feb 11, http://www.shef.ac.uk/news/nr/solar-photovoltaic-pv-spray-painting-lidzey-1.251912

Shell 2013 The New Lens Scenarios, Shell International BV, http://www.shell.com/global/future-energy/scenarios/new-lens-scenarios.html

Solar Power Portal 2012 Self-cleaning coating for solar PV glass could increase efficiency by up to 27 percent, UK Solar Power Portal, http://www.solarpowerportal.co.uk/product_catalogue/self-cleaning_coating_for_solar_pv_glass_could_increase_efficiency_by_up_to

Solar Selections 2012 Kyocera energy management system: Solar Selection website, http://www.solarselections.co.uk/blog/kyocera-energy-management-system-solar-power-battery-storage-solution-introduced-in-japan

Spectrolab 2012 Spectrolab Inc, California, website, http://www.spectrolab.com

Styring S 2012 Artificial photosynthesis for solar fuels *Energy Environ. Sci.*, themed issue on *Artificial Photosynthesis*, *Faraday Discuss.* **155** 357–376

Tesla N 1900 System of Transmission of Electrical Energy, filed September 2, 1897, U.S. Patent No. 645,576, March 20

Tesla N 1914 Apparatus for Transmitting Electrical Energy, filed January 18, 1902, U.S. Patent 1,119,732, December 1

The Engineer 2012 Start-up works on 'spray-on' photovoltaic windows *The Engineer*, June 25, http://www.theengineer.co.uk/energy-and-environment/news/start-up-works-on-spray-on-photovoltaic-windows/1012977.article

Tielrooij K, Song J, Jensen S, Centeno A, Pesquera A, Zurutuza Elorza A, Bonn M, Levitov L and Koppens F 2013 Photoexcitation cascade and multiple hot-carrier generation in graphene *Nat. Phys.*, Feb 24, http://www.nature.com/nphys/journal/vaop/ncurrent/full/nphys2564.html

UDEL 2008 University of Delaware claim 42.8% cell efficiency, with focusing, http://www.udel.edu/PR/UDaily/2008/jul/solar072307.html

Utah 2013 Centre for Acoustic Cooling, University of Utah, http://www.physics.utah.edu/~woolf/acoustics/index.html

Vaughan A 2013 Colourful 'solar glass' means entire buildings can generate clean power *Guardian*, Feb 12, http://www.guardian.co.uk/environment/2013/feb/12/printed-solar-glass-panels-oxford-photovoltaics

Whitlock R 2012 New self-cleaning glass could be used in solar panels *Solar Guide*, http://www.solarguide.co.uk/new-self-cleaning-glass-could-be-used-in-solar-panels

Yirka B 2012 Satellite proposed to send solar power to Earth *Phys. Org.*, April 11, http://phys.org/news/2012-04-satellite-solar-power-earth.html

Yuan J and Branz H 2012 Carrier recombination mechanisms in high surface area nanostructured solar cells by study of 18.2%-efficient black silicon solar cells *Nat. Nanotechnol.* **7** 743–8

Chapter 5

Integration

Integration: grid balancing, energy storage and saving

Many large existing national energy systems are based on the use of large centralised power plants feeding electricity, often over long distances, via distribution networks to consumers, with 'always-on' base-load from continuously run power plants being essential. In the newly emerging system, energy generation would be more decentralised and flexible, at a range of scales, with smart supergrids linking up local as well as remote plants, in a dynamic supply and demand matching network. The base-load concept would be less relevant, indeed in many ways redundant. Having large inflexible always-on plants would be a problem, not a solution, given that the system would have to deal with varying supply, as well as varying demand (REA 2010, Nelder 2012).

The old system was and is very inefficient. Many of the large thermal plants run at 35% efficiency at best, and long-distance AC transmission means that up to 10% of the energy is then lost. The final result is that in the UK about 70% of primary energy is wasted (FoE 2012). Moreover the low efficiency of end-use means that even more is then wasted.

The new system would avoid much of this waste by dynamic matching of (renewable based) supply to demand, less use of thermal plants, careful attention to the efficiency of conversion and of end-use, and by the use of local energy sources wherever possible, combined with high-efficiency, long-distance supergrids, to help balance local variations in supply and demand. This overall approach, it is argued, would help deal with the variability of some of the renewable sources, and should be able to deliver a sustainable and economically viable energy system. In this chapter I look at whether that is realistic.

5.1 Dealing with variability

Energy systems will increasingly have to cope with variable supply inputs, as more renewables come on the grid. The grid system already has to deal with variable demand patterns, and with variable supply inputs (e.g. due to plant failures), and the strategies used to do that can also be extended to deal with variable renewables. For small grid supply perturbations, short-term frequency adjustments can be made, as already happens.

doi:10.1088/978-0-750-31040-6ch5

Larger variations can be backed up using the flexible, fast start-up gas turbines, or plants kept on spinning reserve, that normally deal with demand peaks or sudden plant shut downs. They just have to ramp up and down to full power more often. But there are limits to how much variable supply can be accommodated in these and similar methods, perhaps around 30% of total grid input, depending on the country (IEA 2011a). To go further, more back-up will be needed.

Energy storage is one option. There are new short- to medium-term storage options emerging, which can buffer variable outputs from wind etc, for example various types of flow battery, liquid air storage and even conversion to hydrogen gas via electrolysis. The problem with most storage is that it is inefficient and expensive, much more so than just adding more cheap back-up capacity. So some say the best type of storage is natural gas, stored ready for use when energy is needed from back-up plants. Using fossil gas of course incurs an emission penalty. At some point, as I will be explaining later, that could be avoided by using non-fossil gas in the back-up plants, but for the present the next best option for balancing the grid, and one that can be 100% carbon free, is pumped hydro storage, currently the most widely used large-scale energy storage option.

As I noted in chapter 2, several EU countries have very large hydro capacities, and some of the reservoirs can be used for storage. Norway already has 11 TWh of pumped hydro capacity and might double that. Elsewhere some hydro reservoirs are being converted to pumped storage; for example, Germany's Thuringia State has identified 13 potential sites, including three existing dams, for new pumped storage plants, totalling 5130 MW. Turkey has a large potential for pumped storage (JRC 2013). Even without pumped storage, hydro reservoirs can be used to help balance variable renewables if they have a head of water available. It might be best to keep some of them ready topped up for this, rather than using them for normal grid supply. But that does not help with what is probably going to be perhaps a bigger issue, the regular excess electricity that wind, and also solar, wave and tidal power, will produce when energy demand is low. That is where pumped hydro storage would help

Another possible option is transmission of excess electricity to locations where there are shortages, even if it means long-distance transmission. There would be a need for that in any case, to link to hydro for balancing around the EU. As I have indicated, as well as being used to balance variable renewables around the EU, and for linking to offshore wind in the North Sea and hydro across the continent, high voltage direct current (HVDC) links could also be used for importing solar-derived electricity from the desert areas of North Africa and the Middle East. HVDC supergrid grids can do that with lower losses (2–3%/1000 km) compared with conventional HVAC grid (up to 10%/1000 km).

HVDC has its disadvantages. Expensive transformers/invertors are needed to upload AC from generators and to download DC to AC end-users (assuming that generation and use cannot actually be done with DC), which is why it is best for long distance interconnections. AC is best locally and perhaps also regionally. Interestingly, while backing HVDC for longer-distance links, the SRU, Germany's Advisory Council on the Environment, has suggested that, to avoid the problem with local uplinks, there could be a lower frequency 16.7 Hz overlay AC grid in Europe, at 500 kV. They say that would 'reduce the ratio between line length resistance and frequency, which would represent a threefold reduction relative to today's 50 Hz frequency' (SRU 2011).

All the foregoing is on the supply side. What about managing demand? There is much interest in 'smart grid' or 'dynamic demand' systems, e.g. time-shifting electricity use by incentivising customers to run their energy-using appliances off-peak, through time-of-use tariffs delivered through smart meters. If energy costs more at peak demand times, then consumers might choose to use less then. More aggressively, some loads that are not sensitive to short interruptions in supply can be turned off remotely for a while. Freezer units for example can coast for an hour or two without an electricity input. The technology for 'smart demand' management is developing rapidly. Early tests looked positive (Chen 2004), as have more recent ones (Energi Net 2012a, Deign 2013). The approach is sometimes called 'interactive load management'. Clearly it would only be acceptable if consumers agreed to participate, perhaps in return for lower tariff rates.

A study by consultants Delta says that shaving system 'peaks' reduces the need to use peaking plants which are often less efficient and/or use more polluting fuels; it can also delay or avoid the need for investment in new network capacity. And it can fill system 'valleys', helping to increase and optimise the operation of lower carbon plant. Based on studies by the Brattle Group, they suggest that 'demand response' across all sectors (through the use of time-of-use tariffs) with a 43% take-up, could shift 5% of the system peak to off-peak times (Delta 2011, Faruqui 2007).

Smart grids can also offer other benefits, including diagnostic checks on the performance of heating and ventilation systems, heat pumps, refrigerators, etc with savings of up to 20% being claimed in some cases. In addition, providing consumers with on-line 'smart meter' information about their energy use can help them cut out waste, change their lifestyles and/or lead them to invest in more efficient systems. I will be looking briefly at some energy saving options below and, more radically, at the merits of using heat and green gas rather than just electricity as the main energy supply option, offering new possibilities for storage. After all, heat is much easier to store than electricity and heat stores could be used as a way to balance variable renewables.

Whichever route is taken, some of these options will take longer to deploy than others, and all will be relatively costly. But then so will continuing to rely on fossil fuels, in economic and environmental terms. We have to shift away from them. In addition, some would say nuclear power has no place in this new system. Nuclear plants are usually run 24/7, and so would be of little use for backing-up variable renewables. I will look more at the nuclear issue later, but one thing is clear: it diverts funding from renewables.

The key issue then is: will the new supply and demand management options be enough to allow variable renewables to expand to the 100% that some say is possible at some point? As I indicated in chapter 1, optimists, including many energy analysts, say yes, and have developed detailed scenarios with balancing included; see box 5.1 for some examples.

Pessimists, including some from traditional engineering and/or conservative political backgrounds, say no, renewables cannot be used on a wide scale, especially given their 'intermittency', but offer few alternatives, except maybe nuclear and/or shale gas, perhaps (if they concede that climate issues may be important) with carbon capture and storage added. It will be interesting to see which view prevails. To some extent it may depend on which country is being considered, and by whom!

A recent review of mainly government agency-derived UK energy scenarios to 2050 by the UK Energy Research Centre, perhaps inevitably relayed the standard view that a

Box 5.1. UK and other 2050 energy scenarios for variable renewables

A Poyry consultant's report for the UK government's advisory Committee on Climate Change spelt out how a well integrated and extended grid system could help the UK balance 'stretching but feasible' scenarios with high levels of renewables, up to 94% in their 'Max' electricity scenario. They found that 'the electricity system was able to accommodate these high levels of renewable generation whilst complying with the specified constraints on emissions and security of supply. However, this was at the cost of shedding low variable cost generation and construction of new peaking capacity; predominantly in the two 2050 scenarios and Max scenario'.

Shedding power (or 'curtailing' excess output) is wasteful, but building new low-cost gas-fired capacity is relatively cheap although, they say, not urgent: 'Construction of peaking plant is not required until after 2030 in either the Very High or the High scenario'. Even in their 'Max' scenario, only 21 GW was needed by 2050, although there was some shedding (Poyry 2011a).

A subsequent study of three UK 2050 scenarios by British Pugwash (to which I contributed), made use of the Deparment of Energy and Climate Change (DECC) 2050 Pathways calculator and model. It included a scenario of achieving an 80% renewables input for heat, electricity and transport, in which shedding was avoided by the use of interconnector links to continental Europe, earning the UK £15bn p.a. net from exported excess electricity. Imports via the interconnector links, along with demand management and storage capacity, also helped avoid the need for fossil plant back-up after 2035. The scenario relied on a large offshore wind contribution, along with the use of wave and tidal currents, biomass and biogas, but it avoided all biomass imports. Energy saving was strongly emphasised, along with combined heat and power/district heating (CHP/DH). The overall scenario was slightly cheaper than the rival scenarios in the report, which were based on using nuclear power and gas/CCS. With extensions, and wind-to-gas included, it was claimed that near 100% renewables by 2050 was possible (Pugwash 2013).

A Research Council-supported study of possible UK energy transition pathways by a team of academics was in some ways even more radical than the Pugwash High Renewables scenario, since demand was even more drastically reduced, and local energy projects, including community-scaled CHP, were even more strongly favoured. In one of their scenarios almost all fossil fuel inputs were eliminated by 2050, apart from some carbon capture and storage (CCS) plants (Pearson *et al* 2012).

A report commissioned by the Worldwide Fund for Nature, and produced by energy consultants Garrad Hassan, suggested that renewables could supply 88% of UK electricity by 2030 (WWF 2011). More radically, the Centre for Alternative Technology has produced a series of UK near 100% renewables scenarios aiming to get to zero (net) carbon by 2030 (CAT 2007, 2010, 2013).

Greenpeace has produced studies covering the whole of the EU. For example their 'Battle of the Grids' report, has a near 100% renewables 2050 scenario, with strong demand-side measures, PV at 974 GW, wind 667 GW, bio-energy 336 GW, hydro at 163 GW, concentrating solar power (CSP) at 99 GW, geothermal at 96 GW, and ocean power 66 GW. It was all linked by an EU-wide supergrid, with biomass, geothermal plants and pumped hydro providing back-up (Greenpeace 2011). As I mentioned in chapter 1, several other studies have reached similar conclusions in relation to the EU, and also globally, and dozens of other '100% renewables by 2050' studies have emerged (MNG 2013).

These scenarios all avoid or limit nuclear. However, a nuclear-energy-free future was not something the UK government was considering, according to its chief science adviser Professor Sir John Beddington, at the launch of a long-term UK nuclear strategy, which looked to a possible 75 GW of nuclear by 2050, supplying ~86% of UK electricity (HMG 2013, Carrington 2013).

major nuclear input was needed, but the 'Oceans' global energy scenario from Shell, published at the same time, came to very different conclusions with, in 2050, nuclear making a very small contribution and renewables dominating (UKERC 2013, SHELL 2013). Not covered by the UKERC, a range of other UK studies, reviewed in box 5.1, all come to similar views: very high renewables contributions by 2050 are possible.

5.2 Grid balancing: beyond electricity

While backing gas in the short term, many government plans assume that decarbonised (i.e. non-fossil derived) electricity will be the dominant energy vector in future, with fossil gas being phased out for heating, which would instead be provided by electric-powered heat pumps. However there are other approaches, which have been included in some of the scenarios mentioned above. They see electricity as a poor energy vector, since its transmission involves energy losses and it cannot easily be stored, except via pumped hydro.

By contrast, gas can be easily stored and transmitted. For example, the UK has an extensive low-loss, high transmission efficiency gas grid, which handles around four times more energy than the electricity grid. The gas grid also acts as an energy store, helping the energy system to cope with variable demand. That is crucial, since not only is demand for gas larger than that for electricity, it also varies more (daily) than demand for electricity, as gas-fired heating systems turn on in the evening. Moreover, it does not just have to be fossil gas. Biogas, i.e. biomethane, can be produced from municipal and farm wastes by anaerobic digestion (AD) to provide a carbon-neutral replacement for natural gas.

In addition to its use for heating, some of this 'green gas' could be used for local electricity generation, where needed, in combined cycle gas turbine or fuel cells. Some could even be used for vehicles, as a better option than mostly imported biofuels. So, the argument goes, the focus should be more on a gas grid, increasingly using green gas, moving the emphasis for energy delivery from wires to pipes.

A weak point in this argument is that, in many countries, there probably will not be enough biogas to replace the gas used for heating, much less transport and electricity. As chapter 3 showed, there are land-use constraints to biomass production and there are also limits to how much biowaste is available. There may be a solution. Hydrogen gas can be produced by electrolysis using electricity from excess off-peak wind and other variable renewables, stored or added to the gas main for distribution, or used for electricity production when needed. Some also look to syngas production from renewable electricity for vehicle fuel. All in all, there is much enthusiasm for this 'power-to-gas' idea (Sterner *et al* 2010, Macogaz 2011, Energi Net 2012b, Hydrogenics 2013).

The idea offers some obvious benefits. Excess wind-derived electricity could be converted to hydrogen for use for electricity generation when there was a lull in the wind. There are however some quite severe efficiency loss penalties with some of the energy conversion processes required for making, storing and using 'green hydrogen'. Overall efficiency may only be 30%. However, the technology is improving, and if it is *excess* electricity that is used, which would have been wasted otherwise, that should help the economics. After all, apart from the electrolysis/conversion unit, no new plant has to be built. Germany is taking a lead in 'wind-to-gas' projects for a range of uses; see box 5.2.

Box 5.2. Wind to gas in Germany

By 2013 there were over 7 MW of 'power-to-gas' projects in Germany, using surplus electricity from wind plants to make hydrogen by electrolysis. Car companies were amongst the initial leaders. Audi's product is called 'e-gas', designed for use in their new combustion engines.

However, public energy supply is now also a major focus. ENERTAG is using three 2 MW wind turbines to produce hydrogen which, mixed with biogas made from local corn waste, is used in a CHP/cogen plant, so that electricity can still be fed into the grid when little or no wind is available. It is also feeding its green gas into the natural gas grid, and some of this 'windgas' is being bought by Greenpeace Energy to sell to households (Enertrag 2013).

E.ON has a €5m wind-to-gas pilot project for gas mains injection, while Thüga Group is testing a 360 kW wind-powered electrolysis plant, in Frankfurt. Initially the hydrogen will be injected into the gas grid in a 5% mix, but later it will be reacted with CO_2 from a local power plant to make synthetic methane for full-scale injection. Thüga claim that 'the municipal natural gas network is capable of storing all future generated excess renewable energy', and could be 'the battery of the future'. So they are 'building the battery charger' (ITM 2013, Power in Europe 2013).

An attraction of this approach is that, as well as providing an energy storage medium to balance variable renewables, it can use CO_2 captured from the air to make synfuels. Haldor Topose has developed high-temperature electrolysis/methanisation systems which can use wind-derived electricity to convert water and CO_2 into hydrogen and a range of synfuels/syngases (Hansen 2012a). This approach is also being explored in the UK and Germany by ITM (see box 5.2) and in the EU's CO2RRECT project (CORRECT 2013). In parallel, UK company Airfuel is working on a similar idea, using CO_2 from the air and electricity from wind to make synfuel (Airfuel 2013).

Biogas from AD and other biosources is already being added to the gas grid and this idea is spreading across the EU, with gas from wind adding an extra input, both of the sources replacing fossil gas (Green gas grids 2013). The chief executive of UK company ITM Power, pointed out that 'The UK gas grid is three times the size of the power grid in terms of energy but the main difference is that it has storage—it's relatively easy to store gas. Every day we shift energy from the gas grid into the power grid via power generation. What we need now is a way of moving energy back from the electricity grid into the gas grid'. He added 'all the excess wind we're ever going to produce on the power network can be stored in the gas grid' (Power in Europe 2013).

The hydrogen produced from wind electricity does not necessarily have to be converted to methane. Hydrogen can be injected directly to the gas grid, admixed with methane. Then it would be something like the old town gas (produced from coking coal), which contained methane, hydrogen, and also (poisonous) carbon monoxide. Some use is already made of hythane, which is 20% hydrogen, 80% methane, and higher ratios of hydrogen are possible. There can be issues with embrittlement of metal pipework, although much of that has now been replaced by plastic tubing (Castello *et al* 2005).

Electrolysis is not the only way to generate hydrogen. Most is currently produced by high-temperature steam reformation of methane, although that process also generates carbon dioxide gas. One new idea is to use a low-energy microwave-fired plasma technique instead, with dry carbon rather than CO_2 being a by-product (Gasplas 2013).

Clearly there are many new ideas for new approaches. Putting it all together, in a paper entitled 'Solar-Based Man-Made Carbon Cycle and the Carbon Dioxide Economy', Detlev Möller outlines a visionary plan to link solar electricity production, such as by desert CSP, with CO_2 utilisation via (chemical) air capture (i.e. from the atmosphere), as well as conventional CCS. CO_2 would then be reacted with electrolytically produced hydrogen to produce fuels for direct use or for electricity production when needed. The 'SONNE' approach, as he calls it, would thus seek to build a man-made carbon cycle, like the natural assimilation/respiration carbon cycle by which CO_2 is recycled and changed from waste (emissions) to a resource, all run on solar energy (Möller 2012).

On a more immediate level, there is also a second part to the argument for reducing reliance on electricity transmission and use. Rather than distributing electricity, or for that matter fossil or green gas, to individual domestic consumers for heating, wherever possible heat should be supplied via district heating (DH) networks, fed from high-efficiency community-scaled green-energy-fired CHP plants. DH only makes sense in urban and perhaps suburban areas, but as I noted in chapter 3, biomass and solar-fired DH is now moving ahead across the EU, usually linked to heat stores, and in some cases interseasonal heat stores. So that is an extension of the 'pipe' rather than 'wire' approach, with piped heat as well as piped gas.

None of this means that electricity would not be produced or used. In a way, the 'wire' and 'pipe' approaches are not that different, at least if we are talking, in the pipe version, about large-scale generation of wind-to-green hydrogen. They both rely on having renewable electricity. But they do differ in the main transmission vectors: gas and/or heat in pipes, and electricity in wires. The best balance between heat, gas and electricity and which will, or should, dominate in future is unclear. It will be influenced by the location of the sources and the demand. For example, access for pipes may be hard in some locations. Technological change could also tip the balance of advantage between these vectors. The wind-to-gas route may prove too expensive, whereas the availability of cheap storage of electricity might make electricity more attractive. That of course would also have much wider implications. In the next section I look at the prospects for storage.

5.3 Energy storage

I mentioned earlier that, since energy storage is expensive, it usually made more sense to deal with power grid back-up by using cheap gas turbines. But that may change. Indeed, even without major breakthroughs, there may be a case for storage, when and if there is a large renewable contribution. Dr Tim Fox, Head of Energy and Environment at the UK Institution of Mechanical Engineers, has said: 'For too long we've been reliant on using expensive "back-up" fossil-fuel plants to cope with the inherent intermittency of many renewables. Electricity storage is potentially cleaner and once fully developed is likely to be much cheaper' (Fox 2012).

However, storage will also have to compete with other ways of balancing the grid I have mentioned, such as demand-side management or supergrid links. Given that the various balancing options may interact, it can be hard to identify an optimal mix. Fortunately, the Energy Futures Lab at Imperial College London has produced a 'Strategic Assessment of the Role and Value of Energy Storage Systems in the UK Low Carbon Energy Future'. This adopted a holistic system-wide modelling approach, which helps identify some of the impacts of the interactions (Imperial 2012).

Its headline conclusion was that storage would allow significant savings to be made in generation capacity, interconnection, transmission and distribution networks, and operating costs, and provide up to £10bn of added value in a 2050 UK high renewables scenario. However, it was noted that the relative level and share of the savings changed over time and between different assumptions and scenarios. In the high renewables 'Grassroots pathway' used by the research team, the value of storage increases markedly towards 2030 and further towards 2050, so that carbon constraints for 2030 and 2050 could be met at reduced cost when storage was available. For a bulk storage cost of £50/kW p.a. the optimal volume deployed grew from 2 GW in 2020 to 15 GW in 2030, and 25 GW in 2050. The equivalent system savings increase from a modest £0.12bn per year in 2020 to £2bn in 2030, and could reach over £10bn p.a. in 2050.

As might be expected, the value of storage was highest in pathways with a large share of renewables, where storage could deliver significant operational savings through reducing renewable generation curtailment, i.e. when there was excess wind output available and low demand. In addition, storage could lessen the even larger wind curtailment requirement that would result if there was also a significant amount of inflexible nuclear capacity on the grid. The scenarios with gas back-up plants with added CCS yielded the lowest value for electricity storage: 'adding storage increases the ability of the system to absorb intermittent sources and hence costly CCS plant can be displaced, which leads to very significant savings'.

As can be seen, although it can be very useful in some situations, storage is not a magic solution for all grid-balancing problems. It is best used for specific purposes and durations. Crucially, Imperial say that 'A few hours of storage are sufficient to reduce peak demand and thereby capture significant value. The marginal value for storage durations beyond 6 hours reduces sharply to less than £10/kWh p.a.'.

So it seems we are talking about short storage cycles, ready for the next demand peak, not long-term grid balancing to deal with long lulls in wind availability. That makes sense. Storage is expensive, so the hardware needs to be used regularly to capture excess energy (when it is cheap) and sell it soon after to meet demand peaks, when energy prices are high. This may be fine for short cycles. But how can longer lulls be dealt with, especially in areas where there is a lot of wind capacity? Scottish pumped hydro is one option. Imperial say 'Bulk storage should predominantly be located in Scotland to integrate wind and reduce transmission costs, while distributed storage is best placed in England and Wales to reduce peak loads and support distribution network management'.

The report also offers insights into the interactive nature of the overall system options and operation. As I have indicated, one option for balancing grids in the short term is the use of flexible demand, reducing peaks by time-shifting consumer energy use, for example though interactive smart grids which delay supplies to some loads. Imperial say that 'Flexible demand is the most direct competitor to storage and it could reduce the market for storage by 50%'. So with that, not so much storage would be needed. As I have already mentioned, another option, which might also help with longer-term grid balancing, is the use of interconnectors, including long-distance HVDC supergrid links. There are several advantages. For example, while pumped hydro is the cheapest large-scale bulk electricity storage option, with very high energy storage capacity, not everywhere has good hydro resources, and it can be cheaper to access those areas that do have large pumped hydro storage capacities by using supergrid interconnector links. That could allow those without hydro to export power from wind and so on when there was excess, and import power when there was a long lull in wind availability. As noted earlier, this already happens using the interconnector linking Denmark and Norway.

Interlinks are expensive, but Imperial say that, for the UK, cross-channel links (maybe 12 GW or more) could be 'beneficial for the system because it significantly reduces the amount of curtailed renewable electricity generation in the UK from 29.4 TWh to 15.1 TWh annually'. They added 'this also suggests there will be less scope for storage to be used to reduce the system operating cost through reductions in renewable curtailment. The operating cost savings component is indeed lower in cases with increased interconnection capacity, by about 50% compared to the baseline (Grassroots) case'. So there would not be a need for so much storage.

Even so, in its conclusion, Imperial say that there was a need for perhaps 15 GW of storage, since 'in the Grassroots pathway, storage has a consistently high value across a wide range of scenarios that include interconnection and flexible generation'. So they do see a role for storage.

What exactly is on offer? The Imperial study was primarily about *electricity* storage, but in reality most of the techniques involved do not store electricity as such. More usually, electricity is converted into some other form of more easily stored energy, such as potential energy (pumped hydro). Even batteries convert electric current into stored electrical charge. There are many new energy storage options emerging; it is a rapidly expanding field (Escovale 2013). Box 5.3 looks at some examples of existing and newly emerging electrochemical and electromechanical systems, that is batteries, pumped storage and compressed air storage.

Table 5.1 summarises some characteristics of the storage options looked at in box 5.3. As can be seen, costs vary dramatically, as do energy capacities and power densities. The length of time for which each system can store energy also varies, with pumped hydro and compressed air offering the largest longer term storage options. For more details, see the US Electricity Storage Association (ESA 2013).

There are many other ideas, including systems for converting electricity into heat or (green) gas. That opens up even more interactivity and may also improve the overall efficiency of the system, and perhaps reduce costs. As I have pointed out, it is much easier to store heat or gas than electricity. Heat stores are suited for larger-scale use, including systems using solar- and wind-derived electricity. Hydrogen production and storage systems are also now emerging for use with fuel cells. Shifting to gas and/or heat may open up a range of possibilities for efficient energy storage and use; see box 5.4.

Box 5.3. Energy storage technologies: electrochemical/mechanical options

There are some purely electrical systems, like supercapacitors and magnetic induction devices, which can provide short-term storage, but electrochemical batteries are the most familiar option. However, conventional lead acid types are expensive and bulky and even the more advanced lithium-ion cells, although important for vehicles, are not that suited to bulk energy storage. The largest so far is the Zhangbei project in China, a giant battery array with 36 MWh of output capacity linked to 40 MW of wind and solar plants. Sodium sulfur (NaS) batteries are also used at the multi MW scale. So are nickel cadmium batteries, despite their lower energy density. Metal-air batteries are high energy density and very low cost, but are not directly rechargeable electrically, although some new liquid metal variants might be (Lamonica 2013).

The various types of advanced flow batteries, with round-trip efficiencies of 70% or so, only slightly lower than for lead acid batteries, show promise for larger-scale applications. They mix separate chemical electrolytes to create a charge, in a reversible process (PD Energy 2013). Zinc-bromine and vanadium redox systems are some top contenders, but the US Sandia Lab is looking at electrochemically reversible metal-based ionic liquids, which are non-toxic (Sandia 2012).

In addition to these electrochemical options, there is a range of electromechanical storage technologies. For example, advance flywheels can offer short-term storage and grid balancing possibilities. However, the most developed and widely used approach is pumped hydro storage. As I have indicated, pumped hydro is already used for storing excess electricity from the grid and some non-hydro pumped reservoir schemes (using off-peak power) are also quite large, e.g. there is a 1.87 GW plant on the shore of Lake Michigan (Ludington 2013). Pumped storage, using existing or new hydro reservoirs, or just free-standing pumped storage reservoirs, has significant potential for helping to compensate for the variable output of some renewables.

There are novel pumped storage ideas, such as the Green Power Island proposed off the coast of Denmark, an artificial lagoon built in the sea and linked to a 150 MW offshore wind farm (GPI 2013). A similar idea is being studied in Belgium, a 3 km wide donut-shaped island 3–4 km with a 30 m deep reservoir at its centre and 300 MW of turbine pump and generator units. Excess electricity would be used to pump water out of the reservoir into the sea. When energy demand was high and wind low, the water would be let back into the reservoir through the turbines. Similar ideas have been suggested for tidal lagoons with pumped storage (MacKay 2007).

Another electromechanical option is compressed air storage, for example in large underground reservoirs. In one version of CAES (compressed air energy storage), electricity from wind turbines is used to compress air, which is then stored in caverns underground, for use to supercharge the burning of gas in a conventional turbine (Gaelelectric 2013). In another, compressed air, produced mechanically using electricity from offshore wind turbines, is stored in large inflatable bags, and mounted underwater around the turbine bases for subsequent use in a separate turbine to generate electricity (Pimm and Garvey 2009, Garvey 2011). MIT have come up with an even more ambitious concept, using large hollow concrete spheres mounted in deep water (200 m or more), with water pumped out using energy from floating wind turbines and then let back in through turbines for generation when needed (MIT 2013). Norwegian researchers have come up with a similar idea, claiming 80% round-trip efficiency (SINTEF 2013).

Table 5.1. Energy storage options; key characteristics based on data from ESA 2013.

Pumped storage: very high capacity, low cost, but site specific	$100/kW
Compressed air storage: high capacity, low cost, but site specific	<$100/kW
Flow batteries: high capacity, but low energy density	$100–1000/kW
Metal-air batteries: cheap, very high energy density, but not rechargeable	~$50/kW
NaS, Li-ion, Ni-Cd batteries: expensive, Ni-Cd low energy density	$1 000/kW
Lead-acid batteries: cheaper, but bulky and limited deep cycling life	<$1 000/kW
Flywheels: high power, but low energy density	>$1 000/kW
Capacitors: high efficiency, cheap but low energy density	>$100/kW
High power capacitors: very expensive	<$10 000/kW

Box 5.4. Energy storage technologies: thermal and hydrogen options

Heat is easier to store than electricity, and heat stores can have high-energy storage densities. Hot water can hold about 3.5 times as much energy by volume as natural gas at atmospheric pressure and temperature. However, converting heat back to electricity (by raising steam to drive turbines) can be inefficient so, depending on location and demand, it may be better to use the heat direct. Nevertheless, US IT company Apple are reported to be developing a system which has a tank of water with a mechanical churn driven directly by a wind turbine. Its churning action heats up the water, which is then stored, ready for use for electricity generation when needed.

Another approach is to use excess electricity from wind to heat water in a store via an immersion heater, for use when needed in a district heating network. There is a 200 MW system like this in operation in Denmark. In Scotland, the SHEAP district heating project on the Shetland Isles supplies heat from a waste-to-energy incinerator to 1100 customers, with surplus heat during the night fed to a 12 MWh thermal store. That is now being expanded to link 6.9 MW of wind generation to a 135 MWh capacity heat store with immersion heaters, able to provide 5 days' extra heat (Building4Change 2012).

UK company Isentropic have developed a gravel-filled heat store system linked to a heat pump. They claim the round-trip efficiency is 72–80% (Isentropic 2013). I have already mentioned the use of molten salt heat stores with CSP plants and some new variants are using graphite heat stores (Muirhead 2013). I also noted earlier that there were many hot water solar heat stores linked to DH networks in Denmark, some fed from CHP units. As I indicated, CHP plants can vary the ratio of heat to power output, so that they can be used to balance variable grid power, especially if they also have linked DH/heat stores, with heat being stored when there is excess wind (IEA 2011b).

Any large temperature difference can be used to run a heat engine and there are some systems which use ice as an energy storage medium, although more usually they are part of an air-conditioning/cooling package (BAC 2012, Ice Energy 2013). Developing on that is the idea of cryogenic liquid air storage. UK company Highview Power Storage has demonstrated a 300 kW prototype which stores excess energy at times of low demand by using it to cool air to around minus 190°C via refrigerators, with the resulting liquid air, or cryogen, then being stored in a tank at ambient pressure (1 bar). When electricity is needed, the cryogen is subjected to a pressure of 70 bar and warmed in a heat exchanger.

This produces a high-pressure gas that drives a turbine to generate electricity. The still relatively cold air emerging from the turbine is captured and re-used to make more cryogen. If waste heat from a nearby industrial or power plant is used to re-heat the cryogen, it is claimed the round-trip efficiency rises to around 70% (Highview 2013).

There are also many systems being developed for hydrogen storage, as a gas under pressure or cryogenically as liquid, or chemi-absorbed in metals. In Safe Energy's system, hydrogen is absorbed as metal hydride in a molten mix, which can then be made to release its hydrogen, when required, in a reaction with water, producing hydrogen and heat. That process converts the metal hydride to a metal hydroxide, which can be recycled back to a metal hydride. The magnesium hydride slurry can be stored safely in large quantities at ambient conditions (Safe Hydrogen 2013).

In chapter 3, I mentioned some small domestic-scale examples of hydrogen storage for use with PV solar and fuel cells (see box 3.2). I also mentioned that some hydrogen storage systems have been integrated with electrolytic hydrogen production from wind-derived electricity and hydrogen fuel cells to make a complete green energy system (HARI 2006, Hydrogen Office 2013). There are also microscale fuel cells fed with stored hydrogen, methane or other hydrocarbons, used as an alternative to batteries for laptops and other portable applications (CAP 2008). In addition, many other new hydrogen-based storage options are opening up, some nanotech based (Ozin 2010).

Some argue that storage could be the key to using renewables, and that hardware research priorities should be adjusted accordingly (IMechE 2012). However, while more R&D is sensible, given that many of the newer ideas are untested on a large scale, it is not yet clear what the best mix of systems will be for optimized grid balancing, or the role that storage can and should play (Elliott 2012a).

Other grid-balancing technologies are also moving ahead. For example, a new generation of flexible gas turbines is emerging which can ramp up and down to full power without too much loss of efficiency. GE claim that their new 510 MW FlexEfficiency 50 gas turbines, can achieve 54% efficiency in cycling operations, including 250 starts per year, which is not far below what standard turbines would produce in conventional twice-daily ramp up and down operations, and other suppliers have claimed similar capabilities for their new, more flexible gas turbines (Balling 2011, Probert 2011).

It may be that back-up plants will supersede at least some types of storage, depending on the round-trip energy efficiency of the latter, which may not always be high, even with waste heat recovery; 60–70% is typical, 80% is good. But supergrids might be even better. A US study of offshore wind said 'transmission is far more economically effective than utility-scale electric storage' (Kempton *et al* 2010). Then again, at some locations and scales, storage may be best, offering a perhaps more reliable local source of electricity (or heat), when supplies are low and demand high, than long-distance transmission. After all, there may not always be sufficient input available for the supergrid although, as the next section shows, that depends on the scale, type and location of the renewable capacity.

5.4 A balanced future

Whether using heat, hydrogen, liquid or pressurised air, energy storage is only part of the answer to dealing with renewable variability. Indeed, some analysts think that, in terms of grid balancing, storage might only have to play a small role, given the availability of other balancing techniques (Agora 2013). Moreover, some analysts suggest that, in fact, the 'intermittency' problem is not that great, or at least that dealing with it will not need the deployment of significant new balancing systems for a while. For example, wind energy consultant David Milborrow claims that the UK may have enough back-up plant already in place to cope, and that some fossil-fired plants can actually be retired when wind capacity is added. That depends on the 'capacity credit' of wind, i.e. how much of the wind plant capacity can be relied on statistically to meet peak demand. Milborrow puts the capacity credit of wind at around 30% with low levels of wind on the grid, falling to 15% at high levels (up to 40%). That indicates how much fossil plant can be replaced.

Although Milborrow admits that there will be costs to balancing wind variations, he puts them at around £2.5/MWh at 20% wind, or about £6/MWh at 40% (Milborrow 2009). By contrast, a review of existing studies by the UK Energy Research Centre put the extra costs at £8/MWh at 20% penetration of variable renewables. The UK Committee on Climate Change claimed that the extra cost of dealing with 'intermittency' might be even higher, at £10–20/MWh, for the levels of wind penetration likely to be required to meet the UK's 15% by 2020 renewable energy target, i.e. around 27 GW (CCC 2008).

Certainly, at high levels of renewable input, there will be a need for more balancing, and that could push the cost up, although some measures may be self-financing. For example, given that the UK has a very good wind resource, the cost of importing electricity when needed, via interconnectors, could be more than offset by the sale of exported electricity, when there is excess generated from wind and demand for it elsewhere.

The economics of this type of trade will reflect supply and demand balances. They will vary continually around the network. But the 2013 Pugwash study suggested that the UK could earn £15bn per annum net in this way (Pugwash 2013). A new EU market for balancing energy could open up, with countries like the UK, with good renewable resources, and countries with large hydro reservoirs both being in strong positions.

It is also worth noting that, if there are several different renewables on the grid, then there are some helpful synergies. Wind energy availability patterns are different from those of solar (it is often sunny in the daytime when wind is low) and wave energy is stored wind, while tidal energy is (mostly) unrelated to weather. A Redpoint study for the British Wind Energy Association looked at the optimum balance between wind, wave and tidal. In particular it looked at the extent to which wave and tidal energy could help reduce the grid-balancing costs associated with the use of variable renewables, and also reduce the wind 'spillage', i.e. curtailment when there was too much wind output to be used on the grid. They suggested that, to get the best from the different time correlations of these sources, the optimum might be a 70% wind and 30% wave/tidal current mix or, if tidal range projects were included along with tidal current systems, a 60/40 wind to wave/tidal ratio. Redpoint's RO37 renewables scenario assumes 25 GW of wind capacity by 2020. With the former wind/wave/tidal ratio, the need for fossil-fuelled back-up plants would be reduced by 2.15 GW, with the latter by 2.3 GW, with wholesale costs reduced by up to 3.3%, since there would be less spillage of wind (Redpoint 2009).

The extent of balancing can be improved further by the use of multiple sites for tidal projects. For example, as I noted in chapter 2, given that peak tides occur at different times around the UK coast, a network of tidal current turbines located at different sites around the country could deliver some continuous output. Not all of the output of each site could be matched 100% by others, but it has been estimated that a continuous output of approximately 27% of their maximum collective output could be achieved. Given their wider geographical distribution, if EU tidal stream sites were also added in, the 'firm' output rises to 40% of total collective output (Sankaran Iyer 2011, Waldman 2011).

If small barrages were also included, the percentage would rise further. A study of barrage contributions found that, with 52 mainly very small barrages operating in ebb generation mode, with an overall peak power availability of over 6 GW, the firm power, i.e. continuously available capacity, from the integration would be about 1 GW. However, that was at spring tide periods. It would fall to under 20% of this level during neap tides. A scheme based on 36 larger barrages, including on the Severn and the Wash, could do better and achieve a peak power of 12 GW with a firm power capability of around 4 GW at springs, provided that operation of the barrages included the flood, two-way and variable head systems of generation, although it would be less at neaps (Watson 1994). Including a series of lagoons at geographically dispersed sites around the UK would presumably add to this, but as far as I am aware that has not yet been studied.

A more recent study for the Royal Society looked at small barrage projects in the UK's north west coast, on the Solway Firth, Morecambe Bay, the Mersey and the Dee. If a barrage on the Severn, in the south west, was also included, it was concluded that 'the pulses in electrical output from the north west and the Severn in ebb mode are out of phase' and 'operate in complementary fashion and extend the daily generation window to nearly 20 hours'.

If barrages on the east coast, e.g. on the Wash, Humber, Thames, and perhaps the Forth and Tay, were also included, then, 'partially by reason of their tidal phase lags, a further extension of the daily tidal energy generation "window" might be made' (Yates *et al* 2013).

Pumped storage might add a further time extension during some cycles, but even so, only some of the output from large projects like the Severn Barrage could help with balancing. The very large pulses of energy it produced could mostly not be matched by the outputs of other smaller sites at other times, the large tidal current potential output of the Pentland Firth similarly. Only some of that output could be matched from smaller sites elsewhere when it was not delivering. Even so, depending on which projects are chosen, as can be seen, the spread of geographical tidal phasing offers the option of some firm output. Yates *et al* say a 'combination of tidal stream farms and tidal barrages has the potential to provide continuous base load electricity generation'.

Although this option is only available to countries with tidal resources, most have geographic spreads of other renewables. A test of the viability of using a mix of renewables to balance the national power grid was carried out in Germany in 2006. The Kombikraftwerk 'Combined Power Plant' exercise ran a virtual test, matching measured demand to the real-time output of 36 PV and wind projects backed up by biogas and hydro. The aim was to provide a scale model of how the full national energy system would work. Using one year's data, demand was met with wind and solar contributing around 78% of the total energy. Biogas generators (supplying 17% of the total energy) and pumped-hydro storage capacity (supplying 5%) provided back-up capacity and limited power exchanges were allowed for additional balancing. Averaged over 2006, there was a small net energy gain for export (Kombikraftwerk 2007).

Subsequently, Stadtwerke München, Munich's utility group, also successfully tested a 'virtual power plant' with, in the first stage, 8 MW of cogen (CHP) stations along with 12 GW of renewable plants (a wind farm, and five hydro plants) overseen by a Distributed Energy Management System developed by Siemens.

In a further test, from January 2011 onwards, researchers from the Fraunhofer Institute for Wind Energy and Energy System Technology (IWES) in Kassel ran a new 'Kombikraftwerk 2' virtual power plant trial in the Harz regenerative model region (RegModHarz), using Siemens software, and linking together, via the Internet, 25 plants with a nominal power output of 120 MW and, as simulated storage, a pumped storage power plant and electric vehicles.

Once again the results were good. IWES concluded the combined power plant test showed that, given proper system co-ordination for grid balancing, 'it is technologically possible to let each individual producer feed their electricity into the grid and have the grid remain stable during this process'. It added, 'when many small producers work together then the regional differences regarding wind and sunshine can be balanced out by the power grid or controllable biogas facilities. In addition, surplus power can be stored or turned into thermal energy. A powerful network decentralized, can act as a larger entity' (IWES 2013).

As can be seen, this approach is based on balancing almost totally from internal national sources, with no major imports (or exports) via interconnectors or supergrids. While it clearly can work, SRU, the German Advisory Council on the Environment, has argued that this approach could be sub-optimal. It was better to plan on a wider basis, taking account of resources and facilities in neighbouring countries, e.g. Denmark and Norway (SRU 2011). Certainly, going even wider to a pan-EU supergrid could make balancing even easier by linking up a much wider range of projects in different and much more dispersed geographical locations, including links to pumped hydro capacity.

To test this view, some sophisticated modelling of supply and demand, and then of costs and trade benefits, would be needed. That could ascertain if such a system was economically viable. Some initial energy modelling has been done. TradeWind, a European project funded under the EU's Intelligent Energy-Europe Programme, looked at the maximal and reliable integration of wind power in Trans-European power markets. It used European wind power time series data to calculate the effect of geographical aggregation on wind's contribution to overall generation. It looked ahead to a very large future programme, with its 2020 Medium scenario involving 200 GW of wind capacity, with a 12% pan-EU wind power market penetration (Tradewind 2009).

It found that aggregating wind energy production from multiple countries strongly increased the capacity credit, that is, how much of the wind plant capacity could be relied on to meet peak demand. It also noted that 'load' and wind availability were often positively correlated, improving the system capacity factor, i.e. the degree to which its energy output matches energy demand. For the 2020 Medium scenario, the countries studied showed an average annual wind capacity factor of 23–25%, rising to 30–40%, when considering power production during the 100 highest peak load situations.

In almost all the cases studied, wind generation produced more than average during peak load hours. The report added that since 'the effect of wind power aggregation is the strongest when wind power is shared between all European countries', cross-EU grid links were vital. If no wind energy was exchanged, the capacity credit in Europe was 8%, which corresponded to only16 GW for the assumed 200 GW installed capacity. But since 'the wider the countries are geographically distributed, the higher the resulting capacity credit', if Europe is seen as one wind energy production system and wind energy is distributed across many countries according to individual load profiles, the capacity credit almost doubled to 14%, which it said corresponded to about 27 GW of firm power in the system.

Translating that into reduced back-up plant requirements, another study of EU-wide supergrid links divided the EU up into hexagonal regions of 600 km size (330 km radius), and used hourly data for solar, wind and energy demand. It found that the need for back-up, with 100% renewables supply, could be about halved (Aboumahboub *et al* 2010).

Not everyone agrees with these positive assessments. In a more critical study, Poyry found that improved connectivity would only partially alleviate the volatility of increased renewable energy generation. But it only looked at north west Europe. It left out potentially large inputs from Spain, the eastern EU and, oddly, ignored Ireland (Poyry 2011b).

As I pointed out earlier, mega-schemes like this do raise many issues concerning, for example, the security aspects of the supergrid, and the problems of negotiating grid link access across the whole continent. Balancing supply and demand could also be difficult in practice. One region may have excess electricity available at times when no-one needs it, or there may be a shortfall somewhere when not enough excess is available to trade. More likely there would be regular, hard to manage swings in trading prices, with the prospect of market speculation on stored excess energy opening up yet another dimension. Some people also worry that, if the supergrid system was extended to include North Africa, the EU would be switching from reliance on Middle Eastern oil and Russian gas to North African solar. There is also a danger that EU governments might buy into this approach rather than sorting out energy problems at home. So importing green energy could be used as an excuse for inaction nationally.

Nevertheless, looking on the positive side, a supergrid system could open up a major new energy resource and provide income and wider economic benefits for some relatively poor areas on the periphery of the EU, as long as the trading contracts were fair. There certainly could be some interesting multinational negotiations. For example there are, or at least were, plans for a 3000 km undersea grid from the Algerian town of Adrar, via the island of Sardinia, to mainland Italy, across Switzerland and then to the German city of Aachen, linked to a 150 MW hybrid solar-gas CSP plant at Hassi R'Mel in central M'Zab province, with later expansion expected to 500 MW.

In addition, supergrid links do not necessarily have to be just to the EU. For example there are plans for a US$425m 100 MW CSP plant in Ma'an, in southern Jordan. At present Jordan is sometimes reliant on electricity imports from neighbouring countries. However, surplus production from the CSP plant could be sold to Syria, Egypt and Palestine, whose networks are connected to Jordan.

It seems clear that supergrids could open up not just a major new resource, but also new energy geopolitics. The supergrid and Desertec lobbyists also argue that supergrids are vital for supporting the transition to renewable energy across the EU and elsewhere. They say that 'there can be no transition without transmission' (Friends of the Supergrid 2013).

5.5 Energy conservation and energy efficiency

Many of the issues raised above could be simplified, and the problems reduced dramatically, if energy demand was cut back. Indeed many would say that this should be the first priority. It is argued that saving energy is always cheaper than generating more and that energy saving is the most cost-effective way of reducing emissions.

There are two main aspects to saving energy. Firstly there are technical efficiency measures, which reduce the energy needed to meet some end-use, by more efficient generation and consumption: getting the same utility from less primary energy input. Secondly there are social and behavioural changes, which might avoid waste and may also lead to less actual final energy use. The technical efficiency measures have the attraction of avoiding what might sometimes be perceived as painful and unwelcome lifestyle and social changes, such as being more frugal in the use of energy and energy-based services. Clearly some see this, and talk of low or even zero growth as a threat. However some say Western high consumption lifestyles are in any case neither good nor moral, or indeed sustainable. Even so, while issues like that should be addressed, 'technical fixes' still have some merit. Avoiding waste is surely always sensible. The EU has a target of reducing energy demand by 20% by 2020 and the German energy plan includes a target of a 50% primary energy saving by 2050.

The potential for energy saving via technical measures in modern economies is certainly very large. In the building sector, the technologies should be very familiar, e.g. roof and wall insulation, secondary and/or intelligent glazing and so on. In the industrial sector, process innovation and improved system control can offer significant gains, especially in high energy using industries like aluminum and steel making. The UK Carbon Trust's *Industrial Energy Efficiency Accelerator* looked at fourteen industry sectors, and identified energy, carbon and cost savings averaging 25–30%. Even sizing electric motors properly can be vital, since many are too big for their job (Elliott 2012b).

In all sectors, there are many ways in which small energy losses can be avoided, including losses from equipment left on, or on standby, when not in use. Some of these so-called 'energy vampire' losses can add up to non-trivial amounts nationally.

Overall, looking at key sectors in the UK, a DECC report in 2012 identified a potential for cutting energy demand by 40% by 2030, 155 TWh in all, of which current policy was estimated to capture about 54 TWh (around 35% of total potential); see box 5.5.

The potential, and need, for reducing the energy used by and in buildings is especially large. The International Energy Agency's Energy Technology Perspectives 2012 report, said the building sector was directly or indirectly responsible for about 32% of global energy consumption and for 26% of global total end-use energy-related CO_2 emissions. Moreover it is rising. The International Council of Chemical Association claims that energy use for building heating, cooling, and water heating could rise by almost 60% and greenhouse gas emissions would rise from 3400 million tonnes CO_2 equivalent in 2000 to 5200 million tonnes CO_2 equivalent in 2050 if energy efficiency improvements were not made. However, focusing on Europe, Japan, and the USA, the Association's 'Building Technology Roadmap' claimed that, by 2050, 'tighter new building standards combined with a more ambitious renovation rate could result in a

Box 5.5. UK energy saving options

In 'Capturing the full electricity efficiency potential of the UK', DECC focused on three key sectors, in each case looking at the three largest categories of abatement measures per sector, which together were estimated to deliver ~127 TWh of savings (~80% of total potential) by 2030:

Residential: The top three measures' saving potential is ~58 TWh: CFL lighting, appliances and better insulation, of which ~75% is expected to be captured by current planned policies.

Services: The top three measures' saving potential is ~45 TWh: better insulation, lighting controls and HVAC, of which ~15% should be captured by current/planned policies.

Industrial: The top three measures' saving potential is ~24 TWh: pump, motor and boiler optimisation, of which ~5% is expected to be captured via current/planned policies (DECC 2012).

A report on energy use in buildings from Oxford University claimed that the 477 TWh of gas and oil and 200 TWh of electricity currently consumed in the sector could be reduced to a demand for 100 TWh of renewable electricity supplied by the grid by 2050. As a result, carbon emissions from the sector would then be zero (Boardman 2012).

23% reduction in energy use and GHG compared to 2000', rising to 41% if energy supplies were decarbonised, with net emissions falling by about 70% (ICCA 2012).

A report 'Less is More: Energy Security after Oil', from the UK Association for Environment Conscious Building (AECB), said that investment in energy efficiency can result in major reductions for the least cost, and remove the need for expensive investment in new generating infrastructure. According to the report's principal author, Dr David Olivier, 'Energy efficiency remains as important an opportunity for us as the discovery of a new series of giant oilfields, but without their global warming impact. Many energy efficiency measures save energy worth more than the cost of the measure, so not only do they pay us to save energy, we also save CO_2 at a profit' (AECB 2012).

AECB CEO Andrew Simmonds said 'Efficiency really is the gift that keeps on giving. Efficient use of energy saves on bills now. And it saves the capital cost of all the new extraction, generation and transmission technology that our current levels of energy consumption will demand in the future. We can stick to the cheaper, safer options for new energy, and do without the riskier, pricier ones. None of the energy efficiency measures cited in our report would cost the UK more than about 3p per kWh electricity saved. Who wouldn't want electricity at 3p per kWh, when most consumers currently pay 8–13p per kWh?'

It does seem clear that it is often cheaper to invest in reducing demand than in building new supply capacity. In terms of 'energy return over energy invested' (EROEI), Olivier has estimated that retrofitting solid wall insulation in old buildings may yield an EROEI of up to 50, well above most electricity supply options, large-scale wind perhaps apart (see section 4.3). Moreover, by using electricity more efficiently by adopting low-energy domestic and office electrical equipment, energy-efficient motors, pumps, fans and controls in HVAC, he says that savings as high as 75–90% can often be achieved, with even higher returns on capital than most retrofit insulation (Olivier 2013). In this context it is claimed that it can cost less in the US to replace inefficient appliances with new efficient ones than to buy in energy from low carbon supply (Cary and Benton 2012).

While, as US energy expert Amory Lovins has argued (RMI 2013), large gains can be made, there is the problem that, once the easy and cheap savings have been attained, it may cost more to make further savings. Some energy-saving technologies will get cheaper as the scale of the market for them grows and innovation is stimulated, but once the 'low-hanging fruit' has been picked, getting to higher levels of energy efficiency may be increasingly expensive, with diminishing returns. Moreover there can be some unexpected negative outcomes, which can undermine the savings. While the installation of energy-saving devices and systems may reduce the amount of energy needed to perform some function e.g. to heat a house, it is possible that the householder will then use the money saved to heat the house to a higher temperature.

Of course this is important for those who previously could not afford decent heating levels. But for the rest, there is a risk of overuse. If energy costs less, people tend to use more, if only by being less careful in avoiding wasteful practices (like leaving lights on or doors and windows open). Or they may increase their energy use by buying new appliances like dishwashers or tumble driers. This has been called the 'rebound effect', sometimes also called the 'take-back' effect. Perhaps the worst example, in energy terms, would be if consumers used the money saved by energy conservation to fly on an extra holiday. Then they could wipe out some or all of any carbon emission savings they had made (UKERC 2007, Herring and Sorrell 2009).

There is some debate over the scale of the rebound effect (Chitnis *et al* 2013). Some economists say that the products and services that consumers might spend the saved money on will mostly be less energy intensive than the energy they used for heating, and that the rebound effect may thus only undermine net savings by around 10–15%. But, in general, if any resource becomes cheap, it is likely to be used more. So, all other things being equal, in the context of energy, the end result could be that, although energy efficiency may increase, total energy use (and so emissions) may not reduce much.

The rebound effect is not the only problem. There is also a range of other implementation and uptake problems which may limit the extent to which the large, theoretically possible energy savings can actually be achieved.

Firstly, in relation to new energy-efficient appliances, there is likely to be a delay in deployment, particularly in the domestic sector. Domestic consumers usually only replace appliances occasionally, when they are old or broken, so it would take time to replace the existing range of equipment with more efficient systems.

Secondly, while over their lifetime the more efficient appliances will cost consumers less to run, they may cost more to buy, and this can be a disincentive for those with limited budgets. The old-age pensioner in a damp, draughty high-rise flat may only be able to afford a cheap one-bar electric fire.

Thirdly, house owners who, due to pressure on jobs and careers, tend these days to move home frequently, may not think it worthwhile to retrofit energy-saving measures like insulation in their home. The payback times may be too long for them to benefit.

Fourthly, there can be technical problems. Houses designed with good insulation and airtight double/secondary glazed widows can be stuffy and may need (powered) ventilation. Ventilation or dehumidifiers may also have to be provided to avoid condensation. In some areas there can be a build-up of mildly radioactive radon gas inside, emitted from rocks under the building, in which case powered ventilation may have to be provided.

In some situations it may also be more economic to opt for supply rather than savings. For example it may be cheaper and easier to link a high-rise residential building to a DH network than to retrofit external wall cladding to reduce energy loss. A UK study comparing CHP/DH with domestic insulation even concluded that, for a typical late 1960s/early 1970s London house, in a terrace of five houses, connection to DH gave a lower capital cost per tonne of CO_2 displaced than alternative insulation measures (Orchard Partners 2010).

However, in general, it is not a matter of choosing between 'renewables' or 'efficiency'. Usually they are complementary. Both are needed and both need to be deployed as rapidly as possible. Despite the potential problems with consumer take-up and with the rebound effect, energy/fuel conservation is vital for any strategy for cutting emissions. If demand can be cut back, for example by good 'low-energy' building design, with built-in passive and active energy-saving measures in new buildings, as well as retrofitted upgrades to existing buildings where viable, it becomes easier for energy supply technologies to meet it in environmentally sound ways, via the use of renewables.

On this view, conservation and renewables complement each other. Indeed, if combined they can help reduce the rebound effect. For example, if the money a householder saves from investing in energy efficiency is used to buy in energy from renewable sources, then the emission savings will be captured, rather than lost, as they would be if this money had been spent on (fossil) energy-intensive goods and services.

How can consumers be stimulated to adopt more efficient ways of using energy? Increased direct energy pricing, via an energy or carbon tax, is an obvious approach. However, this is socially regressive, impacting most on the less well off. An arguably less brutal approach is via personal carbon rationing. Consumers would be allocated carbon credits to limit their overall consumption of energy and fuel. These credits would be tradable, so that consumers who managed to use less could sell any excess to those who were less frugal, with a market for credits being created. The annual allocations for everyone would be gradually reduced.

There has been a long debate on the pros and cons of personal carbon rationing (Parag and Strickland 2009, Bird and Lockwood 2009). Enthusiasts say, not unreasonably, that personal rationing schemes would have an immense educational value, making people very aware of their carbon debts, leading hopefully to behavioural change. But some people may see personal carbon rationing as an unwarranted imposition, for example requiring invasive policing, given the potential for evasion and abuse.

In the worst case, there could actually be a net increase in emissions. The rich and energy profligate would simply buy in credits from the poor, to escape the overall cap limits, while given their high value to the rich, the poor would be tempted to sell their credits and try to buy in dirty 'off list' energy as a replacement. Of course, in the best case, at least for the climate, given the increasing cost of energy and credits, the rich may cut back to some extent, and/or invest in efficiency/self-generation, while the poor may decide to do without some energy services, in order to continue to sell off their credits. So in that case there may be some reduction in emissions. But at what social cost?

For some critics, it is not clear that operating at the individual consumer level, via personal caps and rationing, is the best way ahead. It is far easier, they claim, to put emission caps on the relatively few energy generation and supply companies, although they will then pass on costs to consumers, unless they are somehow constrained.

One idea is to impose a carbon tax on energy companies, but ring-fence the money raised to ensure that it is invested in green energy projects. Targeted 'hypothecation' of money like this could allay fears that the money would just disappear into company cash flows. More radically, funds from a national 'carbon fee' imposed on suppliers could be given directly to consumers to invest in their own energy projects (Hansen 2012b). There might be problems with that, not least given the rebound effect since, in an open market, energy-saving measures may not ensure actual energy/emission reduction.

What is needed, it might be argued, is for money that consumers save through energy efficiency to be invested in renewable supply. Then the rebound effect is avoided and all the carbon savings captured. Carbon taxes/fees may not achieve that, but personal carbon rationing might not either. As an alternative, a 2012 report by the UK Green Alliance and WWF-UK argued for an energy efficiency feed-in tariff (FiT) to create a market for 'negawatts', i.e. saved energy (Cary and Benton 2012).

Some attempts have been made in the UK to link energy policies and uptake support schemes for energy saving and renewables, with grant/FiT support schemes for installing domestic energy supply microgen units being conditional on having achieved specified levels of energy saving. An energy efficiency FiT might be linked into that. But it is still some way off, as are many other potentially worthwhile energy saving and demand management ideas (Whitehead 2012).

Overall, there is a way to go before the large potential of energy saving can be attained. Despite the cost saving opportunities, energy saving and efficiency is still often given a low priority. For now, fuel is still relatively cheap and there are powerful vested interests driving its continued use and, beyond a few simple and easy measures, the public's appetite for adjusting lifestyles to reduce energy waste seems to be relatively limited.

5.6 A sustainable energy future: costs and policy choices

A serious commitment to energy efficiency, coupled with accelerated development of renewables and efficient end-use applications like DH, could arguably lead to a balanced and sustainable energy supply and demand system. The cost of making the transition will be large, but perhaps not much larger than the cost of replacing the existing range of energy technology as that becomes obsolete. For example, over the next couple of decades, most existing nuclear plants around the world will have to be replaced, having reached the end of their operating lifetimes. The same is true for coal and gas plants.

Some argue that switching over to renewables and reducing energy wastage will lead to a net cost saving over time, not least since, with most renewables, there are no fuel costs, while the costs of fossil fuels are bound to continue to increase. In addition, if the use of fossil fuels is not eliminated, there will be ever-rising social and economic costs due to the impacts of climate change. Stern suggested that they could be up to 20% of global Gross National Product, while the cost of ameliorating climate change would be much less, perhaps 2% of GNP (Stern 2007).

These social and environmental costs are not usually reflected fully, or sometimes not at all, in estimates of generation cost, although some assessments may add in 'carbon costs' based on the impact of the various carbon taxes in place or planned. This is not an economics textbook, or a study of climate policy, so I will not go (far) into the often very complex and controversial issues linked to long-term cost and price estimation. But clearly there are inevitably many variables and unknowns when attempting to estimate likely costs for future energy systems. For example, there is no way to predict what interest rates (for borrowed money) will be in the future, or what discount rates (for investment returns) will be used over the payback period.

Nevertheless, economic assessments are made, and below I present some of the results of one carried out in the UK. This uses 'levelised' cost estimates. Put (very) simply, the initial capital cost is spread over the lifetime of the plant's operation, with a sum added to the 'technical' generating cost, using a discount rate reflecting the average cost of capital. Quite apart from not knowing what the actual cost of capital will be over long periods, this is an obvious simplification. In reality, for some plants, the initial capital costs may be paid back early on, so that the plant then runs at lower (just operational) costs.

An added complexity is that large capital-intensive plants may take up to a decade to build, during which time they will not be paying back any costs from sales of energy. To cope with that, economists use a figure for what they call 'overnight costs', as if the plant could be built instantly, overnight, avoiding interest payments on borrowed capital during construction. Comparisons made on that basis will artificially favour large long-lead-time projects but, as I have indicated, the levelised cost approach also has problems which can make it hard to use to compare technologies of very different types, lifetimes and levels of development. For example, who knows what fuel costs will be in the future?

Despite all these problems, decision makers do use data like that in table 5.2 below, and I have made use of some in chapter 2. It certainly offers interesting predictions.

Table 5.2. Generation cost estimates per MWh delivered in the UK based on Mott MacDonald data; levelised costs in £/MWh (Mott MacDonald 2011).

Electricity option	Current cost	Cost in 2040	Lowest estimate
On land wind	83–90	52–55	(51)
Offshore wind	169	69–82	(60)
Tidal barrage	518	271–312	(120)
Tidal stream	293	100–140	(97)
Wave (fixed)	368	115–140	
(floating)	600	200–300	(175)
Hydro (small/run of river)	69	52–58	(44)
Photovoltaics (PV)	343–378	60–90	(43)
Biomass (wastes/SRC)	100–171	100–150	(80)
Biogas (AD/wastes)	51–73	45–70	(32)
Geothermal	159	80	(55)
Nuclear (PWR/BWR)	96–98	51–66	(39)
Gas-CCS	100–105	100–105	(95)
Coal-CCS	145–158	130	(110)

The above data are mostly taken from the executive summary of the Mott MacDonald report (e.g. figure 2, for current costs), but other estimates (in chapters 3 and 7) are also offered, based on different assumptions, e.g. about progress down learning curves, some of which, at the bottom of the ranges, yield lower costs; see bracketed figures. The report has some inconsistency: at one point current nuclear costs are put at £89/MWh and those for 2040 offshore wind as £100–130/MWh (perhaps on the assumption that floating wind systems are not successful), but in the above table, those figures apart, to be conservative, I have used their higher estimates.

As can be seen, currently, on land wind does better than all other options, hydro and biogas apart, although by 2040 the cost range for nuclear overlaps with all but biogas, the latter including sewage/landfill gas, some of the cheapest current energy sources. Although they have relatively small resource bases, other AD biogas options may well prove to have larger resources and be equally competitive, hence the low costs quoted.

This assessment was produced by a major consultancy company, Mott MacDonald, for the UK government's Advisory Committee on Climate Change. Mott MacDonald did admit to having been 'bullish' in their estimates of nuclear costs, which were evidently based on industry projections. Certainly its low 2040 figure for nuclear, of £39/MWh, seems somewhat unlikely, especially for, it has to be said, as yet unbuilt and untested reactors. Indeed, even the higher central estimates have been challenged as being too low (Harris *et al* 2012). Critics have also seen the estimates for future costs of some renewables as unduly pessimistic, given that many were now operating at full scale, and overall some were very critical of this analysis (No2Nuclear Power 2011).

There are many other such assessments, some of which come to different conclusions, some being more favourable to renewables, sometimes using different accounting frameworks (IEA 2013, Prysma 2013). I will be looking at some more UK examples in the next chapter. But Mott MacDonald's figures should at least indicate that, even on a perhaps limited basis of comparison, some renewables are competitive and more will become so in the UK.

For comparison, a subsequent global assessment of 'Renewable Power Generation Costs in 2012', from IRENA, the International Renewable Energy Agency, put the current levelised cost of energy for renewables as shown in table 5.3. As can be seen, the ranges for current renewable energy costs are all lower than those offered by Mott MacDonald, in some cases significantly so, and IRENA predicts continuing reductions in future.

Table 5.3. Global levelised cost of energy in 2012, in 2011 US$ (IRENA 2013) UK £/MWh conversions added to aid comparison with table 5.2, at 2013 exchange rates.

New small hydro	$0.03–0.07/kWh	£20–40/MWh (rounded up)
Large hydro	$0.03–0.06/kWh	£20–40/MWh
Biomass (non OECD)	$0.05–0.06/kWh	£30–40/MWh
Geothermal	$0.05–0.09/kWh	£30–60/MWh
On land wind	$0.08–0.12/kWh	£50–80/MWh
CPV	$0.15–0.31/kWh	£100–200/MWh
CSP	$0.22–0.25/kWh	£140–160/MWh

These are weighted average figures across regions. In some areas they can be lower. For example IRENA notes that in North America wind costs can be $0.04–0.05/kWh, making wind cheaper than gas. In all cases (hydro and geothermal apart) they predict 10–20% or more cost reductions by 2020.

Disputes over the economics of the various energy systems will no doubt continue, and there certainly is a need to continue to reduce costs, so that a wider range of renewables become economically viable. Some may never be so, but it is perhaps premature to judge, given the relatively early stage in the transition to new technology. It may not be possible, or even wise, to 'let a hundred flowers bloom' and then pick the best, but foreclosing options at too early a stage to limit risk can in fact be a risky strategy. In the 1970s, the UK government dismissed wind energy as unlikely to be significant and in the 1980s it also shut down research on wave and tidal energy and on solar energy and deep geothermal, decisions that now look unfortunate (Elliott 1997, 2010a). That said, given inevitably limited funding, choices do have to be made, and the strategic issues are not easy. For example, should the focus be on the short-term costs or the long-term potential scale of the resource? As I will be illustrating in the next chapter, choices about how much and how quickly to develop specific options can be politically difficult.

Leaving aside the economic and technical problems, there are also many other problems ahead, including institutional resistance and public opposition to some of the new ideas. Making large-scale changes will involve significant social and political as well as technical challenges and choices. In this situation, what can be portrayed as simple technical fixes are usually more popular than anything that sounds like it may require major changes. Even arguably quite risky options, like nuclear power, may be seen as better than facing what for many is the unknown, the use of renewable energy.

Avoiding change is not an option. Whatever happens, current energy systems will have to be replaced, as the existing plants reach the end of their useful lives, and moving away from fossil fuel use seems essential. Some say that we should replace existing nuclear plants with new nuclear plants and keep the overall energy system more or less the same. Others say nuclear can and should be expanded dramatically. For example there have been proposals for the UK to expand its nuclear contribution by almost a factor of ten to 90 GW by 2050 (Smith School 2012). Even more dramatically, the Russian government is aiming for nuclear to supply 45–50% of its electricity by 2050, and 70–80% by 2100.

Major expansions on this scale might be seen as unrealistic for most countries, and indeed overall, given the limited and diminishing global uranium reserves (EWG 2006, Dittmar 2013). By contrast, renewables have no problems of fuel scarcity, or rising fuel costs, no waste disposal issues, and no weapons proliferation risks (Elliott 2010b).

Some say we could have both nuclear and renewables. But on any significant scale, nuclear and renewables are operationally incompatible, highlighting a key system integration issue. If there is a large amount of basically inflexible nuclear capacity on the grid, it gets in the way of dealing with variable renewables. Nuclear plants are usually run 24/7 to recoup their large capital cost. Wind plants have no fuel costs, so there is an incentive to run them whenever there is wind. So, for example, which gives way when energy demand is low and/or if there is plenty of wind? It does not make sense to curtail either output but, in the absence of large amounts of storage or export links, that is what would have to happen, undermining the economics of whichever was curtailed.

It is possible to vary the output from nuclear plants to some extent. Indeed this is already done to a small degree in France (and was done in Germany) to partially meet the daily energy demand cycles (Bruynooghe 2010). But cycling from high to low power more often would impose larger cost and, possibly, safety penalties. In addition to the extra thermal stress, cycling reactors up and down to full power generates contaminating radioactive by-products, which interfere with operation and take time to be dispersed safely. It seems unlikely that cycling could be done rapidly or frequently enough to balance the hour-by-hour varying output from wind and other variable renewables. EDF have said that the new European Pressurised Water Reactor 'can ramp up at 5% of its maximum output per minute, but this is from 25% to 100% capacity and is limited to a maximum of 2 cycles per day and 100 cycles a year. Higher levels of cycling are possible but this is limited to 60% to 100% of capacity' (EDF 2008). So it would be of little help in balancing renewables on a day-to-day basis.

Although it is conceivable that new technologies or operational opportunities (e.g. storage or exporting excesses via supergrids) might help resolve integration issues like this, some critics argue that the nuclear option is also fraught with other difficulties, both in the short and long term (e.g. fuel scarcity and a long-term radioactive waste legacy). At best, they say, it offers an expensive interim supply option; at worst it could divert resources from the long-term development of renewables (Sovacool 2011). Fossil fuel CCS is also sometimes seen in the same way, as an expensive interim supply option, diverting attention from renewables and leaving behind a long-term legacy of waste (stored CO_2).

We urgently need to decide which way to go. The basic issue seems irrefutable. The use of fossil fuel must be phased out. But choices have to be made on its replacement. In the next chapter, I will be looking at what choices are being made around the world.

Summary points

- The **newly emerging energy system** will need new ways of integrating and balancing variable supply and demand, and there are several options.
- **Energy storage** may be needed but flexible cheap **back-up plants** using green fuel may win out.
- **Wind to gas**—using excess wind (or PV) energy to make hydrogen for generating power when there is a lull in wind, is a clever new idea, if it is economically viable.
- **Supergrids** can play a major role, helping to balance local/regional/national variations while **smart grids** can help manage demand interactively.
- **Energy saving** measures would make everything easier, but are not always easy to deploy successfully.
- **Nuclear power** just gets in the way; it is not compatible with the new flexible system.
- Some renewables are already **competitive** and more should be soon.
- **Choices** have to be made.

References

Aboumahboub T, Schaber K, Tzscheutschler P and Hamacher T 2010 Optimization of the Utilization of Renewable Energy Sources in the Electricity Sector, Recent Advances in Energy and Environment Conference, http://www.wseas.us/e-library/conferences/2010/Cambridge/EE/EE-29.pdf

AECB 2012 Less is More, UK Association for Environment Conscious Building, http://www.aecb.net/publications/less-is-more-energy-after-oil/

Agora 2013 12 Insights on Germany's Energiewende, Agora Energiewende, a joint initiative of the Mercator and European Climate Foundations, http://www.agora-energiewende.de/fileadmin/downloads/publikationen/Agora_12_Insights_on_Germanys_Energiewende_web.pdf

Airfuel 2013 Airfuel company website: http://www.airfuelsynthesis.com and http://www.imeche.org/news/archives/12-10-15/UK_engineers_create_petrol_from_air.aspx

BAC 2012 Barcelona District Cooling with BAC's Ice Thermal Storage, Baltimore Aircoil Company http://www.baltimoreaircoil.com/english/14032/barcelona-district-cooling-with-bacs-ice-thermal-storage

Balling L 2011 Fast cycling and rapid start-up: new generation of plants achieves impressive results, Modern Power Systems, Jan reprint, http://www.energy.siemens.com/fi/pool/hq/power-generation/power-plants/gas-fired-power-plants/combined-cycle-powerplants/Fast_cycling_and_rapid_start-up_US.pdf

Bird J and Lockwood M 2009 Plan B? The prospects for personal carbon trading, Institute for Public Policy Research, London IPPR Report, http://www.ippr.org/publicationsandreports/publication.asp?id=696

Boardman B 2012 Achieving Zero, Environmental Change Institute, Oxford University, www.eci.ox.ac.uk/research/energy/achievingzero/

Bruynooghe C, Eriksson A and Fulli G 2010 Load-following operating mode at Nuclear Power Plants (NPPs) and incidence on Operation and Maintenance (O&M) costs. Compatibility with wind power variability, European Commission Joint Research Centre Institute for Energy, http://tinyurl.com/6n7grs9

Building4Change 2012 Storing heat energy makes the most of fluctuating wind power, BRE Trust, March 27, http://www.building4change.com/page.jsp?id=1219

CAP 2008 Energy Storage Technologies: A Comparison, CAP XX review, http://www.cap-xx.com/resources/reviews/strge_cmprsn.htm

Carrington D 2013 Nuclear-free future not an option for UK energy strategy, says chief advise *The Guardian*, March 26, http://www.guardian.co.uk/environment/2013/mar/26/nuclear-free-future-energy-strategy

Cary R and Benton D 2012 Creating a market for electricity savings: Paying for energy efficiency through the Energy Bill, Green Alliance/WWF, Oct, http://assets.wwf.org.uk/downloads/creating_a_market_for_electricity_savings_oct_2012.pdf

Castello P, Tzimas E, Moretto P and Peteves S 2005 Techno-economic assessment of hydrogen transmission and distribution systems in Europe in the medium and long term, The Institute for Energy, Report EUR 21586 EN, EC Joint Research Centre, Petten, The Netherlands

CAT 2007, 2010, 2013 Zero Carbon Britain, Centre for Alternative Technology, Machynlleth, first, second and third editions, http://www.zerocarbonbritain.com/

CCC 2008 Building a low-carbon economy—the UK's contribution to tackling climate change, The UK government's advisory Committee on Climate Change, London, 185, http://www.theccc.org.uk/publication/building-a-low-carbon-economy-the-uks-contribution-to-tackling-climate-change-2/

Chen A 2004 Multi-building Internet demand-response control system: the first successful test *University of California, Berkeley Research News* **510** 486–4210, http://www.lbl.gov/Science-Articles/Archive/EETD-demand-response.html

Chitnis M, Sorrell S, Druckerman A, Firth S and Jackson T 2013 Turning lights into flights: Estimating direct and indirect rebound effects for UK households *Energy Policy* **55** April, 234–50

CORRECT 2013 Programme website, http://co2chem.co.uk/carbon-utilisation/co2rrect

DECC 2012 Capturing the full electricity efficiency potential of the UK, Department of Energy and Climate Change, London, http://webarchive.nationalarchives.gov.uk/20121217150421/, http://www.decc.gov.uk/en/content/cms/emissions/edr/edr.aspx

Deign J 2013 Demand response: Faroe Islands wind project serves valuable lessons *Smart Grid Update*, Jan 9, http://us.smartgridupdate.com/fc_fcbi1lz/lz.aspx?p1=0594277S2223&CC=&p=1&cID=0&cValue=1

Delta 2011 Driving a Resource Efficient Power Generation Sector in Europe, Consultants report on pricing, http://www.delta-ee.com/images/downloads/pdfs/2011/Delta%20Final%20Report%20-%20Driving%20a%20Resource%20Efficient%20Power%20Generation%20Sector%20in%20Europe.pdf

Dittmar M 2013 The end of cheap uranium *Sci. Total Environ.*, May 16, http://www.sciencedirect.com/science/article/pii/S0048969713004579

EDF 2008 EDF's submission to the UK governments renewable energy strategy consultation: 'UK Renewable Energy Strategy: Analysis of Consultation Responses' Prepared for: Dept of Energy and Climate Change Log Number 00439e, http://www.berr.gov.uk/files/file50119.pdf

Elliott D 1997 Renewables Past, Present and Future, NATTA Report, Network for Alternative Technology and Technology Assessment, Milton Keynes

Elliott D (ed) 2010a *Sustainable Energy* (Basingstoke: Palgrave Macmillan) pp 67–68

Elliott D (ed) 2010b *Nuclear or Not?* (Basingstoke: Palgrave Macmillan)

Elliott D 2012a Energy Storage *Renew Your Energy* blog, Environmental Research Web, July 16, http://environmentalresearchweb.org/blog/2012/06/energy-storage.html

Elliott D 2012b Energy in Industry *Renew Your Energy* blog, Environmental Research Web, Jan14, http://environmentalresearchweb.org/blog/2012/01/energy-in-industry.html

Enetrag 2013 Company website, http://www.enertrag.com/en/project-development/hybrid-power-plant.html

EnergiNet 2012a Uncontrollable wind can be controlled *EnergiNet Denmark*, March 8, http://energinet.dk/EN/FORSKNING/Nyheder/Sider/Den-ustyrlige-vind-kan-styres.aspx

EnergiNet 2012b Dutch, Danish and Belgian green gas link up *EnergiNet Denmark*, http://www.energinet.dk/EN/GAS/Nyheder/Sider/100CO2neutral.aspx

ESA 2013 Electricity Storage Association website, http://www.electricitystorage.org/

Escovale 2013 A leading energy storage consultant's website. See their annual energy storage technology reviews, http://www.escovale.com/

EWG 2006 Uranium Resources and Nuclear Energy. Energy Watch Group, Berlin EWG-Paper No 1/06, http://www.energywatchgroup.org/Uran.60+M5d637b1e38d.0.html

Faruqui A 2007 The Economics of Dynamic Pricing for the mass market, Battle group report, http://www.brattle.com/_documents/UploadLibrary/Upload578.pdf

FoE 2012 Up in Smoke, Friends of the Earth UK briefing note, http://www.foe.co.uk/resource/briefing_notes/iib_elec_graphic.pdf

Fox T 2012 Comments at the launch of an Institution of Mechanical Engineers policy statement on Energy Storage, May, http://www.imeche.org/knowledge/policy/energy/policy/ElectricityStoragePolicy Statement

Friends of the Supergrid 2013 Friends of the Supergrid trade lobby group, http://www.friendsofthe-supergrid.eu

Garvey S 2011 The dynamics of integrated compressed air renewable energy systems *Renewable Energy* **39** (1) 271–92

Gaelelecric 2013 CAES system, http://www.gaelectric.ie/index.php/energy-storage/

GasPlas 2013 Norwegian CO_2-free hydrogen production technology, http://www.gasplas.com/w3/

Green gas grids 2013 Green gas grids, Intelligent Energy Europe, http://www.greengasgrids.eu/

Greenpeace 2011 Battle of the Grids, http://www.greenpeace.org/international/Global/international/publications/climate/2011/battle of the grids.pdf

GPI 2013 Green Power Island website, http://www.greenpowerisland.dk

Hansen J B 2012a Biogas Upgrading and High Temperature Electrolysis, Training the Trainers, Haldor Topsoe, Copenhagen

Hansen J 2012b Climate change is happening now—a carbon price must follow *The Guardian*, Nov 29, London, http://www.guardian.co.uk/environment/2012/nov/29/climate-change-carbon-price?intcmp1/4122/

HARI 2006 Hydrogen and Renewables Integration project, Prof Tony Marmont, West Beacon Farm/ Centre for Renewable Energy Systems Technology, Loughborough University, UK, http://www.ieahia.org/pdfs/HARI.pdf

Harris G, Heptonstall P, Gross R and Handley D 2012 Cost estimates for nuclear power in the UK, ICEPT Working Paper WP/2012/014, Imperial College, London, August

Herring H and Sorrell S (ed) 2009 *Energy Efficiency and Sustainable Consumption* (Basingstoke: Palgrave Macmillan)

Highview 2013 Highview Cryogenic energy storage, company website, http://www.highview-power.com/ *Also see* www.imeche.org/news/press-release/12-02-01/Liquid_air_the_solution_to_wind_power_s_unreliability.aspx

HMG 2013 The UK's Nuclear Future, HM Government, www.gov.uk/government/organisations/department-for-business-innovation-skills/series/nuclear-industrial-strategy

Hydrogen Office 2013 Wind-Hydrogen project, Fife, Scotland, http://www.hydrogenoffice.com/

Hydrogenics 2013 Wind to gas: Hydogenics company website, http://www.hydrogenics.com/

ICCA 2012 Building Technology Roadmap, International Council of Chemical Association, Brussesl, http://www.icca-chem.org/

Ice Energy 2013 Ice Energy company website, http://www.ice-energy.com/

IEA 2011a Harnessing Variable Renewables: a Guide to the Balancing Challenge, International Energy Agency, Paris, http://www.iea.org/publications/freepublications/publication/name,34724,en.html

IEA 2011b Co-Generation and Renewables: Solutions for a Low-Carbon Energy Future, International Energy Agency, Paris, http://www.iea.org/publications/freepublications/publication/name,3980,en.html

IEA 2013 Medium Term Renewable Energy Market Report 2013, International Energy Agency, Paris, www.iea.org/w/bookshop/add.aspx?id=453

IMechE 2012 Policy statement on Energy Storage, Institution of Mechanical Engineers, London, May, http://www.imeche.org/knowledge/policy/energy/policy/ElectricityStoragePolicy Statement

Imperial 2012 Strategic Assessment of the Role and Value of Energy Storage Systems in the UK Low Carbon Energy Future, Energy Futures Lab, Imperial College, London, http://www3.imperial.ac.uk/newsandeventspggrp/imperialcollege/administration/energyfutureslab/newssummary/news_5-7-2012-14-8-41

IRENA 2013 Renewable Power Generation Costs in 2012: An Overview, International Renewable Energy Agency, Abu Dhabi. Summary at http://www.irena.org/menu/index.aspx?mnu=Subcat&PriMenuID=36&CatID=141&SubcatID=261

Isentropic 2013 Isentropic gravel heat storage system, company website, http://www.isentropic.co.uk/

ITM 2013 First Sale of 'Power-to-Gas' Plant in Germany, ITM Power, company press release, March 13, http://www.itm-power.com/news-item/first-sale-of-power-to-gas-plant-in-germany

IWES 2013 The virtual power plant – stable supply of electricity from renewable energies, Fraunhofer Institute for Wind Energy and Energy System Technology (IWES), Kassel, Research News, March 26, http://www.fraunhofer.de/en/press/research-news/2013/march/the-virtual-power-plant.html

JRC 2013 Assessment of the European potential for PHS, European Commission Joint Research Centre, http://setis.ec.europa.eu/newsroom-items-folder/jrc-report-european-potential-pumped-hydropower-energy-storage

Kempton W, Pimentaa F, Dana E Verona D and Colle B 2010 Electric power from offshore wind via synoptic-scale interconnection, PNAS, http://www.pnas.org/content/early/2010/03/29/0909075107.full.pdf+html

Kombikraftwerk 2007 Combined Power Plant test, Institute for Solar Energy Supply Systems (ISET) at the University of Kassel, http://www.kombikraftwerk.de/index.php?id=27S

LaMonica M 2013 Ambri's Better Grid Battery, MIT Technology Review, Feb 18, http://www.technologyreview.com/featuredstory/511081/ambris-better-grid-battery/

Ludington 2013 Pumped storage plant on lake Michigan, http://www.consumersenergy.com/content.aspx?id=1830

Mackay D 2007 Enhancing Electrical Supply by Pumped Storage in Tidal Lagoons, http://www.inference.phy.cam.ac.uk/sustainable/book/tex/Lagoons.pdf

Macogaz 2011 Power to gas, Macogaz fact sheet, http://www.gasnaturally.eu/uploads/Modules/Publications/marcogaz_power2gas_fact_sheet.pdf

Milborrow D 2009 Wind Power: Managing Variability, Greenpeace UK, www.greenpeace.org.uk/media/reports/wind-power-managing-variability

MIT 2013 Wind power even without the wind, MIT News release, April 25, http://web.mit.edu/newsoffice/2013/wind-power-even-without-the-wind-0425.html

MNG 2013 List of 60 or more reports on the feasibility and costs of decarbonising the UK, Europe and the world, MNG NGO network, UK, http://www.mng.org.uk/gh/scenarios.htm

Möller D 2012 Solar-based man-made carbon cycle and the carbon dioxide economy *AMBIO* **41** 413–19

Mott MacDonald 2011 Costs of low-carbon technologies, report for the Committee on Climate Change, May, http://hmccc.s3.amazonaws.com/Renewables%20Review/MML%20final%20report%20for%20CCC%209%20may%202011.pdf

Muirhead J 2013 Tracking Graphite storage's progress *CSP Today*, April 4, http://social.csptoday.com/technology/tracking-graphite-storage%E2%80%99s-progress

Nelder C 2012 Why baseload is doomed *Smart Planet*, March 28, http://www.smartplanet.com/blog/energy-futurist/why-baseload-power-is-doomed/445

No2NuclearPower 2011 The Cost of Nuclear Power, No2NucearPower, Briefing, Feb, http://www.no2nuclearpower.org.uk/reports/EconomicsBriefing.pdf

Olivier D 2013 All 'Sources' of Energy are Not Equal, Conference presentation, http://www.claverton-energy.com/all-sources-of-energy-are-not-equal-presentation.html?utm_source=feedburner&utm_medium=email&utm_campaign=Feed%3A+ClavertonEnergyGroup+%28Claverton+Group+News+Summary%29 Also downloadable from www.gmp.uk.com/index.php?option=content&task=view&id=593

Orchard Partners 2010 Retrofit for Future, Orchard Partners report for the UK Technology Strategy Board, http://www.orchardpartners.co.uk/ Summary: http://www.claverton-energy.com/technology-strategy-board-retrofit-for-future-a-study-to-minimise-co2-emissions-for-typical-uk-housing-comparing-combined-heat-and-power-district-heating-with-insulation-march-2011.html

Ozin G 2010 What Can Nanochemistry do for Hydrogen Storage? *Materials View*, Nov 24, http://www.materialsviews.com/what-can-nanochemistry-do-for-hydrogen-storage/

Parag Y and Strickland D 2009 Personal Carbon Budgeting, ECI/UKERC report, http://www.eci.ox.ac.uk/research/energy/downloads/paragstrickland09pcbudget.pdf

PD Energy 2013 Vanadium Flow Cells company website, http://www.pdenergy.com

Pearson P, Hammond G, Leach M and Foxon T 2012 Transition Pathways to a Low Carbon Electricity System, Final Dissemination Conference, Power Point summary, http://www.lowcarbonpathways.org.uk/lowcarbon/conference/A_Short_Guide_to_UK_Transition_Pathways.pdf

Pimm A and Garvey S 2009 Analysis of flexible fabric structures for large-scale subsea compressed air energy storage *J. Phys.: Conf. Ser.* **181** 012049

Power in Europe 2013 ITM Power: accelerating power-to-gas *Power in Europe* **64** (8) April 1, http://www.itm-power.com/wp-content/uploads/2013/04/Platts-April13.pdf

Poyry 2011a Analysing technical constraints on renewable generation to 2050, Poyry consultants report to the Committee on Climate Change, March

Poyry 2011b North European Wind and Solar Intermittency Study, Poyry consultants report, http://www.poyry.com/news-events/news/groundbreaking-study-impact-wind-and-solar-generation-electricity-markets-north

Probert T 2011 Fast starts and flexibility: Let the gas turbine battle commence *Power Engineering International*, June 1, http://www.powerengineeringint.com/articles/print/volume-19/issue-6/features/fast-starts-and-flexibility-let-the-gas-turbine-battle-commence.html

Prysma 2013 RE-COST Study on Cost and Business Comparisons of Renewable vs. Non-renewable Technologies, Prysma consultants report for the IEA, http://iea-retd.org/archives/publications/re-cost

Pugwash 2013 Pathways to 2050: Three possible UK energy strategies, British Pugwash, London, http://www.britishpugwash.org/recent_pubs.htm

REA 2010 Renewable Energies and Base Load Power Plants: Are They Compatible?, German Renewable Energies Agency report, http://www.unendlich-viel-energie.de/en/details/article/523/renewable-energies-and-base-load-power-plants-are-they-compatible.html

Redpoint 2009 The benefits of marine technologies with a diversified renewable mix, Redpoint consultants report for British Wind Energy Association, London

RMI 2013 Reinventing Fire, Rocky Mountain Institute, Colorado, http://www.rmi.org/ReinventingFire

Safe Hydrogen 2013 Company website, http://www.safehydrogen.com

Sandia 2012 Sandia National Laboratories researchers find energy storage 'solutions' in MetILs, Press release, https://share.sandia.gov/news/resources/news_releases/metils/

Sankaran Iyer A, Couch S, Harrison G and Wallace A 2011 Phasing of tidal current energy around the UK and potential contribution to electricity generation, University of Edinburgh paper, http://www.supergen-networks.org.uk/filebyid/630/file.pdf

Shell 2013 The New Lens Scenarios, Shell International BV, http://www.shell.com/global/future-energy/scenarios/new-lens-scenarios.html

SINTEF 2013 Storage power plant on the seabed *Science Daily*, May 15, http://www.sciencedaily.com/releases/2013/05/130515085343.htm

Smith School 2012 Towards a low carbon pathway for the UK, the Smith School of Enterprise and the Environment, University of Oxford, http://www.smithschool.ox.ac.uk/

Sovacool B 2011 *Contesting the Future of Nuclear Power* (Singapore: World Scientific)

SRU 2011 Pathways towards a 100% renewable electricity system', SRU, German Advisory Council on the Environment, Berlin, http://www.umweltrat.de/SharedDocs/Downloads/EN/02_Special_Reports/2011_10_Special_Report_Pathways_renewables.html

Stern N 2007 The Economics of Climate Change, report for UK Treasury, http://webarchive.nationalarchives.gov.uk/+/http://www.hm-treasury.gov.uk/independent_reviews/stern_review_economics_climate_change/sternreview_index.cfm

Sterner M, Pape C, Saint-Drenan Y-M, von Oehsen A, Specht M, Zuberbuhler U and Sturmer B 2010 Towards 100% renewables and beyond power: The possibility of wind to generate renewable fuels and materials, Fraunhoffer Institute/IWES, http://www.iwes.fraunhofer.de/de/publikationen/uebersicht/2010/towards_100_renewablesandbeyondpowerthepossibilityofwindtogenera.html

Tradewind 2009 Tradewind study of wind power integration in the EU, co-ordinated by the EWEA http://www.trade-wind.eu/

UKERC 2007 The Rebound Effect, UK Energy Research Centre, London, http://www.ukerc.ac.uk/support/tiki-index.php?page=0710ReboundEffects

UKERC 2013 The UK energy system in 2050: Comparing Low-Carbon, Resilient Scenarios, UK Energy Research Centre, London, www.ukerc.ac.uk/support/tiki-download_file.php?fileId=2976

Waldman S 2011 Phase differences between EU tidal stream sites and their effect on the variability of the total resource (summary), http://www.firecloud.org.uk/linkedin_pubs/110925_diss_summary.pdf

Watson W 1994 Firm Power, Network for Alternative Technology and Technology Assessment (NATA), Milton Keynes and also Renew 90, NATTA, 1994

Whitehead A 2012 Demand-side measures and the case for decapacity payments *Energy* blog, http://www.alan-whitehead.org.uk/pdf/decapacitypayments.pdf

WWF 2011 Positive Energy: how renewable electricity can transform the UK by 2030, World Wide Fund for Nature, London, http://assets.wwf.org.uk/downloads/positive_energy_final_designed.pdf

Yates N, Walkington I, Burrows R and Wolf J 2013 Appraising the extractable tidal energy resources of the UK's western coastal waters *Phil. Trans. R. Soc.* A **371** 1471–2962

Chapter 6

Policy

Policy: global review and strategic development issues

Renewables are expanding rapidly around the world. As noted by REN21, the global Renewable Energy Network, in 2011 renewable sources supplied 16.7% of global final energy consumption, with investment in renewables increasing by 17% to a record $257bn, 94% higher than the total in 2007, the year before the world financial crisis. Globally, renewables accounted for over 71% of total new electricity capacity additions in 2011, raising renewables' total share of electricity capacity to over 31%. That year also saw photovoltaic module prices drop by 50% and on land wind turbines by close to 10%, bringing the price of the leading renewable power technologies closer to grid parity with fossil fuels such as coal and gas (REN21 2012).

As I have indicated, expansion has subsequently continued. In 2012 wind capacity reached over 280 GW, largely driven by China, while solar PV reached 100 GW globally. Moreover, as the technologies develop, there are significant prospects for further expansion. In a report on 'Deploying Renewables' the International Energy Agency said that 'a portfolio of renewable energy technologies is becoming cost-competitive in an increasingly broad range of circumstances, in some cases providing investment opportunities without the need for specific economic support'. It included established hydro, geothermal and bio-energy in the list. It added that 'cost reductions in critical technologies, such as wind and solar, are set to continue' (IEA 2013).

6.1 Prospects and problems

From the above it would seem that the prospects for renewables are very good. However, as previous chapters have shown, there are problems, some related to variability in output and land-use requirements, but some related to challenges from rival energy options. Gas is one, coal another, but nuclear, as a non-fossil option, is perhaps the main rival.

The Global Energy Assessment which, as I mentioned in chapter 1, was produced by an international team led by the International Institute for Applied Systems Analysis, included a series of possible low-carbon pathways. One achieved maximum demand

doi:10.1088/978-0-750-31040-6ch6

saving through a focus on energy efficiency, and had a strong emphasis on renewables and carbon capture and storage (CCS). In that scenario, nuclear power was progressively phased out in most of the world, and then totally by 2050. Although it also included scenarios with nuclear expansion, the GEA saw '*nuclear energy as a choice, not a requirement*' (GEA 2012).

Following the major nuclear accident at Fukushima in Japan in 2011, several countries evidently agreed with this view (notably Germany, Italy, Belgium and Switzerland) and so joined those who had already abandoned nuclear power earlier (e.g. after the Chernobyl disaster in 1986) or had avoided going down the nuclear route in the first place and had focused instead on renewables and energy efficiency, including, within the EU, Austria, Denmark and Ireland. However, some countries continue to support nuclear power, notably China, India, the USA, the UK, Russia and South Korea. In some of them (but not all), renewables are often emphasized less. Certainly there is a strong imbalance in subsidy funding around the world, with nuclear getting the lion's share (see table 6.1). In some countries the historical difference is stark; in the USA, the nuclear industry benefited from a cumulative $185.38bn in federal subsidies between 1947 and 1999, while renewables, starting later, received $5.93bn between 1994 and 2009.

The nuclear issue is only one of many influencing the success or otherwise of renewables, but it is, I feel, an important factor. Global R&D funding for nuclear still dwarfs that for renewables, and the simple fact is that a dollar spent on nuclear is a dollar not spent on renewables. If there were unlimited funds and technical resources, it might be possible to develop both, but that is not the situation, so choices have to be made.

Some argue that abandoning nuclear would reduce diversity of energy supply. Certainly there is value in diversity, in spreading risks, but actually renewables are not just a single option, but a wide range of very different technologies. If diversity is important then, arguably, renewables are the best bet, spreading risks widely. By contrast, nuclear is based, so far, on just one approach, which has had over 40 years of extensive funding, and it needs even more. Keeping nuclear in the mix may therefore

Table 6.1. Government-funded energy subsidies in IEA countries (millions of US$ (1997)).

Technology	1974–2007 Cumulative %		1998–2007 Cumulative %	
	Total	Share	Total	Share
Energy efficiency	38 422	8.9	14 983	14.2
Fossil fuels	55 072	12.8	11 114	10.6
Renewable energy	37 333	8.7	10 709	10.2
Nuclear fission/fusion	236 528	54.8	43 667	41.5
Hydrogen and fuel cells	2 824	0.7	2 824	2.7
Transmission and storage	15 717	3.6	5 388	5.1
Other	45 204	10.5	16.599	15.8
Total	**430 855**	**100**	**105 194**	**100**

Source: Sovacool (2011). (IEA countries = most of the major ones.)

even reduce diversity. Moreover if, as some claim likely, nuclear continues to display a 'negative learning curve' with capital costs continually rising (Gruber 2010, Harris *et al* 2012), and with costs for renewables likely to keep falling, we would have backed the wrong horse.

As I mentioned in chapter 4, nuclear fusion is an even longer shot, decades away at best (Harris 2013), and yet it demands an even larger stake. Around $20bn has already been spent globally and even more is now being allocated to it. Oddly, the payoff may not only be uncertain but also relatively small. At best, the UK Atomic Energy Authority has suggested that, even if all goes well with the current extensive and very costly R&D efforts, fusion might only be able to supply around 20% of global electricity by 2100 (UKAEA 2007). There may be breakthroughs in fusion research, but we either have to rely on that or wait until 2100 to perhaps get 20%. Renewables supply nearly that amount now, and can make an immediate and expanding contribution to responding to climate change and energy security threats. In effect, part of the case for renewables is that we already have a free, safely working nuclear fusion reactor in the sky, and we can use it now, given the necessary funding to continue to develop effective energy conversion technologies.

The case for renewables is strengthened by the fact that they are also popular. It seems clear that, around the world, most people support most renewables strongly and oppose nuclear power, with the level of opposition to nuclear being very high in some countries, up to 80% or more. Table 6.2 provides a global overview. Within the EU, 70% supported renewables as being the top energy priority in a major 2012 poll, as against 18% backing nuclear. In Portugal, Austria, Spain, German and Denmark, support for renewables was at over 80% (Eurobarometer 2013).

I do not want to dwell on the nuclear issue, not least since I have written about that elsewhere (Elliott 2010, 2012a), so in what follows I have provided a snapshot of the state of play globally, region by region, focusing mainly on the progress made on renewables, but also indicating policies on nuclear power where relevant. This is not a fully up-to-date or comprehensive survey, but may give a feel for how renewables are progressing around the world. For updates see the annual reviews produced by REN21 (REN21 2012).

Table 6.2. Global public support for renewables.

Solar power	97%
Wind power	93%
Hydro power	91%
Natural gas	80%
Coal	48%
Nuclear	38%

Source: From a June 2011 IPSOS public opinion survey in 24 countries: Argentina, Australia, Belgium, Brazil, Canada, China, France, UK, Germany, Hungary, India, Indonesia, Italy, Japan, Mexico, Poland, Russia, Saudi Arabia, South Africa, South Korea, Spain, Sweden, Turkey and the USA (IPSOS 2011).

6.2 Europe

Europe has taken a lead in developing renewable energy, in part due to its policies on mitigating climate change. By 2011 the EU was meeting 13% of its gross energy consumption from renewables (Europa 2013), and it seems set to reach its target of 20% by 2020, along with a commitment to reducing energy use by 20% by 2020.

As table 6.3 shows, within the overall 20% energy target, there are some striking differences in the aspirations and achievements so far of the EU member countries. In what follows I will try to explain how that has occurred, starting with the early movers.

Demark was one of the early pioneers, following its decision in 1985 not to develop nuclear power. Wind now meets over 25% of its annual electricity requirements, and there are plans to expand that, and other renewables, as well as energy efficiency.

Denmark aims to be zero carbon by 2050. Its interim plan to 2020 aims to get to a 35% renewable share of final energy consumption, with electricity use being met almost entirely by wind. The 2050 policy envisages the use of all fossil fuels being phased out, including for transport, with renewables taking over (Danish Government 2011/12).

Austria decided not to start up its newly built nuclear plant following Chernobyl in 1986, and has pushed ahead strongly with renewables, building on its large hydro base. By 1997 it was generating 72.7% of its electricity from renewables, about half from hydro, and aiming for over 78%. Grass roots initiatives made it a solar heat collector leader and its 2020 renewable energy target (which it should reach) is one of the EU's highest, at 34%.

Germany, which is the EU's largest economy, had already been pushing ahead strongly with renewables. Following Fukushima, it shut eight nuclear plants and decided to phase out the rest by 2022. To replace them it has accelerated its renewables programme. It initially aimed to get at least 35% of its electricity from renewables by 2020, but has raised that target to 40%, thereafter expanding in stages to 80% by 2050 (Maue 2012).

On land, wind has been expanding rapidly. It reached 30 GW by 2012. Next, more than 20 offshore 'wind parks' have been approved in the North Sea and three more in the Baltic, all outside the 12 nautical mile Exclusive Economic Zone (EEZ). Inside the EEZ, four wind parks have been approved in the North Sea and two in the Baltic. Photovoltaic (PV) is also continuing to expand. It had reached 30 GW by 2012 (Fraunhofer 2012).

Longer term, the viability of the new German energy system will depend significantly on whether it can upgrade and balance its grid system. For the moment it is using fossil fuels to provide most of the backup but, as I have noted earlier, it is looking to biogas and to wind-to-gas projects to play a major role. A study carried out by DENA, the Deutsche Energie-Agentur GmbH (the German Energy Agency) says that, on current plans, by 2050 there would be 240 GW on the grid, with 170 GW of renewable and 61 GW of fossil-fired plants (DENA 2012). They would presumably have to be CCS linked to avoid carbon emissions, although some could be biomass/biogas fired.

The development of storage capacity is another key issue. The government has allocated €200m to 60 energy storage research projects. Energy efficiency is also a major focus. The aim is to cut primary energy consumption by 20% by 2020 and then in stages to 50% in 2050, with electricity use falling by 10% and 25% in 2020 and 2050 respectively. The potential does seem to be there for moving to 80% renewables and

Table 6.3. EU renewable energy use and 2020 targets.

	% of gross energy consumption (Eurobserver 2012)	
	2010	2020 agreed targets
Sweden	47.6	49
Latvia	33.1	40
Finland	33.0	38
Austria	30.9	34
Portugal	26.8	32
Estonia	25.6	25
Romania	24.1	24
Denmark	23.5	30
Slovenia	18.5	25
Lithuania	18.3	23
Spain	15.1	20
France	13.3	23
Bulgaria	12.8	16
Germany	12.3	18
Greece	11.2	18
Italy	11.2	17
Poland	10.6	15
Czech Republic	10.4	13
Slovakia	9.5	14
Hungary	8.2	13
Ireland	6.1	16
Cyprus	6.0	13
Belgium	5.6	13
Netherlands	4.4	14
United Kingdom	3.8	15
Luxembourg	2.8	11
Malta	0.4	10
EU overall 2020 energy target		20

The data above are for total energy use, that is for heat, transport fuels and electricity. Some EU countries benefit from large existing hydro contributions, notably Austria, Finland, France, Portugal, Spain and Sweden, which perhaps makes comparisons unfair. To put that issue in perspective, it may be helpful to look at the EU's 2010 *electricity* targets, which were set both with and without hydro. Austria's 78.1% 2010 target with hydro was only 21.1% without, Sweden's target fell from 60% to 15.7%, Finland's from 35% to 21.7%, Portugal's from 45.6% to 21.5%, Spain's from 29.4% to 17.5%, while France's fell from 21.1% to 8.9%. For the rest the changes were small, and the overall ranking order did not change very significantly with hydro excluded. For example, the UK, Malta and Luxembourg stayed at the bottom, with the UK's 10% target reduced to 9.3% without hydro. The same pattern emerges in the *energy* data above.

indeed beyond, to 100% (UBA 2010, SRU 2011). This is not just to generate electricity. The plan is for renewables to achieve an 18% share of gross final energy consumption by 2020, a 30% share by 2030, 45% by 2040 and 60% by 2050 (Vertrung 2013). To get there, strong community and grass roots involvement is envisaged and, as I will show later, that seems to be emerging.

The energy transition is thus seen as being social as well as technological (Ethics Commission 2011). So far overall progress has been quite good. Although the use of fossil fuel has risen slightly, despite the nuclear phase-out programme, Germany has continued to export power, helping out France during the 2011 winter cold spell. Moreover, with renewables building up, its greenhouse gas emissions declined 2.1% in 2011 (Reuters 2012), although they rose slightly in 2012, partly due to a cold winter.

Germany is not alone in the EU in planning a nuclear phase-out and a rapid expansion of renewables. Following Fukushima, **Italy** voted in a referendum to abandon plans for new nuclear (with 94% opposing nuclear), and is pushing ahead with renewables. It has been a strong player in PV solar (15 GW so far) and also wind (8 GW so far). After Fukushima, **Belgium** also decided to phase out all its nuclear plants, as did **Switzerland** and both are backing renewables strongly. **Portugal** remains non-nuclear. It already gets over 50% of its electricity from renewables with, in addition to hydro and some large PV projects, 4.5 GW of on land wind capacity in place. It has also been developing its large wave and offshore wind resource. But the global recession and Euro crisis hit it hard and there has been a slowdown. Similarly for non-nuclear **Greece**, although it is looking to inward investment in renewables, PV especially, to help it recover. Strongly anti-nuclear **Ireland** is developing its very large wind resource, with 2 GW installed so far, and is also looking at wave and tidal power. It may also export electricity from wind to the UK.

Spain had a long-established nuclear phase-out policy and has been amongst the leaders in developing renewables. By 2012 it had installed nearly 23 GW of wind, and over 4 GW of PV. It also has around 2 GW of concentrating solar power (CSP) in place. But a change of government in 2012 led to a slowdown of both the nuclear phase-out and of the renewables programmes, and with the recession hitting Spain particularly hard, in 2012 the feed-in tariff (FiT) support system was temporarily suspended. **Sweden** introduced a nuclear phase-out policy after Chernobyl, but has now softened it, although it is a leader in renewables, given its large hydro and biomass resources. **Norway** is similarly blessed, but has resisted the nuclear option.

The swing to the left in **France** opened up the possibility of at least a partial nuclear phase-out and a new renewables expansion programme. With opposition to nuclear growing (to around 70% in some polls), the Socialist Party was elected in 2012 on a promise to consider reducing reliance on nuclear power from 70% (of electricity) to 50% by 2025, with renewables filling the gap. It initiated a public debate on this proposal, with a final decision expected in late 2013.

The situation in the **United Kingdom** is very different. The Conservative–Liberal Democrat coalition government has backed a large private sector-led nuclear expansion (HMG 2013). The UK currently gets just over 17% of its electricity from nuclear. Despite having the EU's best renewable resources, and good public support, the UK renewables programme is unambitious by EU standards, with an energy target of just 15% by 2020. The use of a competitive market-based 'Renewables Obligation' support system, rather than FiTs, has arguably not helped (Elliott 2012b). One bright spot is its offshore wind programme, nearing 3 GW by the end of 2012, with plans for up to 18 GW by 2020. The UK also has some pioneering wave and tidal current technology (DECC 2012).

The devolved government of **Scotland** has much more radical plans. It is opposed to new nuclear and by 2012 got over 36% of its electricity from renewables (compared to around 11% for the UK as whole), with its large wind resource playing a major role. It is aiming to get 100% of its electricity use matched from renewables by 2020, well ahead of any country in the EU, Denmark aside. Scotland also plans to get 30% of its total energy from renewables by 2020, and by 2030 it wants to be 'largely decarbonised', and to have phased out nuclear (Scottish Government 2011, 2012a, 2012b). In 2014 it will vote on whether to become independent of the UK. That raises some energy policy issues (Elliott 2013a).

Some of the new EU member countries are doing very well, Latvia especially, but also Estonia, Romania, Slovenia and Lithuania (see table 6.3). In 2012, Lithuania decided not to replace its Soviet-era Ignalina nuclear plant, which had been closed as part of its EU accession agreement. However some of the other new EU members are still wedded to nuclear, **Finland** especially. It is expanding its programme, despite having a very large renewable energy resource and a high renewables target. Some of the ex-Soviet Bloc countries in eastern and central Europe, now in the EU, are similarly keen to retain and even expand nuclear, although after Fukushima, Bulgaria abandoned plans for new plant.

Beyond the EU, **Russia** is strongly focused on exporting gas for the short term and on nuclear as its main future option. This is despite the fact that nuclear is strongly opposed by a clear majority of the public. A BBC Globespan poll found that, after Fukushima, opposition to nuclear had risen from 61% (in 2005) to 80%, with 43% opposing nuclear outright and 37% opposing new nuclear plants, and just 9% being in favour (BBC 2011). Russia has very large potential renewable energy resources. The wind resource in north west Siberia and northern Russia has been put at 350 GW. The current plan, however, is to get just 4.5% of Russia's electricity from renewables by 2020 (IES 2010).

A similar lack of interest also seems to apply, so far, to the governments of many of the ex-Soviet Bloc states to the east of the EU, some of which have very large renewable energy resources. For example, the wind resource in **Kazakhstan** has been put at 210 GW. It could be that, at some point, these huge resources will be tapped, possibly for exporting power to the EU, or to China, opening up a new area of geopolitics. For the moment, the emphasis in the region is on oil and gas, or gas transit pipeline services. **Turkey** similarly is focusing on gas transit and, longer term, despite its very large wind potential (perhaps 20 GW), is planning a nuclear plant, although it is also aiming to get 30% of its electricity from renewables by 2023, using wind, solar and hydro.

Some countries in the **Middle East** are also looking to nuclear, including Saudi Arabia, the UAE, Jordan and Egypt. However, as I indicated earlier (see box 3.3), some are also now looking to solar, an obvious resource for them (GTM 2013a). Interestingly, after Fukushima, Kuwait and Bahrain decided to abandon their nuclear plans. In general, with large solar projects proposed, including Saudi Arabia's plan for 41 GW by 2032 and the UAE's ambitious Masdar solar city initiative, and Qatar aiming to get 20% of its electricity from renewables by 2030, renewables should lift off seriously in this area (Menasol 2013).

6.3 Asia and the Pacific area

Following the Fukushima nuclear disaster in 2011, all of **Japan's** nuclear plants were closed down for tests, and Japan is developing an alternative approach to energy supply. It is based on energy efficiency and a major commitment to renewables, including a new, quite generous FiT for PV solar, a major offshore wind programme and more support for other marine renewables (Elliott 2012a, 2012c).

Offshore projects obviously make sense in a country where land is at a premium. They include, very symbolically, a proposed 1 GW floating wind turbine array off the coast from Fukushima. Although final plans have yet to be confirmed, it has been said that overall around 8 GW of wind capacity might be in place by 2030, which seems minimal. The wind lobby has talked of 25 GW on land and 25 GW offshore, and they put the longer-term wind potential at over 200 GW, on land and offshore, even taking account of locational constraints, similar to Japan's current total generating capacity (JWPA 2010).

While the future for wind is being thrashed out, the government has decided to allow geothermal energy projects in newly opened areas of national parks. It is claimed that this could result in the development of up to 2 GW of capacity by the 2020s. As a start, again very symbolically, a 500 kW geothermal plant is to be installed at a hot spring in Fukushima City. New biomass projects have also been started, including algae for biofuels. Even more is expected of PV, given the generous new FiT. In Germany PV has reached 30 GW using that approach. PV could do even better in Japan, given its generally more favourable climate. The Ministry of Environment has even predicted that PV could eventually reach 100–150 GW, with a projected value of $9.6bn.

For the next phase, a 2030 energy plan is being produced. A consultation report in 2012, 'Towards a strategy—where public debate is pointing', was based on the three options that the government had put forward: zero nuclear, 15% nuclear or 20–25% nuclear, with renewables taking up most of the slack, at 30–35%. Energy use would also be reduced.

With 100 000-strong demonstrations against nuclear occurring regularly and weekly rallies outside the Prime Minister's residence, opposition to nuclear was clearly very strong. In one poll 47% opted for zero nuclear, 16% went for 15%, while 13% chose 20–25%. In September 2012 the government announced that it would aim to get to zero nuclear 'in the 2030s' (Japan Times 2012).

Given that Japan, the world's third-largest economy, was originally planning to expand nuclear from its pre-Fukushima 26% to 45% of total electricity by 2030, a shift to zero would be very significant, and would involve major changes, especially since the aim is still to reduce its greenhouse gas emissions by 23–25% by 2030 from 1990 levels. However Japan's then industry minister, Yukio Edano, said: 'I don't think the zero scenario is negative for Japan's economy. On the contrary, it can create growth as efforts to develop renewable energy and improve energy-efficiency could boost domestic demand' (Edano 2012).

A detailed 2030 strategy has yet to emerge, and with an election in late 2012 leading to a new more pro-nuclear government, it may be delayed or modified but, although one plant has been restarted and a few more might be, it is hard to see how substantial changes in policy could be made given the strength of public opinion.

Interestingly, the emergency energy-saving measures imposed by the government after Fukushima cut peak summer energy use by between 10 and 20% depending on the area, partly as a result of behavioural changes. Consumers were exhorted to reduce demand so as to avoid blackouts, with electronic billboards at metro stations and screen displays with TV newscasts showing current electricity use. There were still some blackouts; however, the campaign did show what could be done. For example in the area covered by Tepco, small-business users cut usage by 18%, households by 17% and heavy industry by 15% (Froggatt *et al* 2012). In parallel, on the technology front, there are some clever new ideas emerging for 'smart grid' integrated domestic energy management systems.

With its track record in technological innovation, in addition to deploying renewables on a large scale, Japan may yet show us all how to develop more sustainable ways of living.

China is pushing on with renewables very significantly, with these and other non-fossil fuel options expected to provide around 15% of its total energy needs by 2020. The nuclear programme is a small part of that, aiming to get to 4% of electricity by 2020. Renewables already supply 17%. The nuclear development progamme was suspended following Fukushima, but restarted in late 2012, although on a reduced basis, with the government approving a 'small number' of projects in each of the next five years. An IPSOS poll found that 58% of the Chinese public opposed nuclear after Fukushima (IPSOS 2011).

Wind power is the new big thing. There is said to be over 75 GW of capacity installed so far, well ahead of every other country. And that is just the start. The Chinese Wind Power Development Roadmap 2050 stipulates that China will have 200 GW installed wind capacity by 2020, 400 GW by 2030, and 1000 GW by 2050 (Qi 2011).

However, China is trying to refocus what has so far been something of an uncontrolled boom with, for example, insufficient attention having been paid to providing the necessary grid links. The result has been that, although China had over 42 GW of wind capacity installed by the start of 2011, only an estimated 31 GW was grid-linked. Many of these projects, most of which were in remote areas in the north west and poorly served by grid links, were often unable to dispatch their full potential output to users, most of whom are in the major urban areas on the coast. This issue is now being addressed. The 12th Five-Year Plan (2011–15), which calls for 100 GW of capacity by 2015, includes significant investment in grid infrastructure (Ma and Fu 2011, Modern Power Systems 2011).

Hydro has been the big, more conventional, renewable option for China and, given the remote location of giant projects like the 18.2 GW (soon to be 22.4 GW) Three Gorges dam, it too has grid implications, although more progress has been made there. A series of high voltage direct current (HVDC) links have been built to east and south China, over distances of around 1000 km, to transfer electricity from the Three Gorges hydro plant. The total capacity of the HVDC links is 7200 MW, with line losses put at about 3%.

On a very different technology scale, PV solar is the other big new renewable option. So far China has focused on exporting PV, becoming a world leader, but now it is looking to large-scale deployment. Its earlier target of 10 GW by 2020 has now been upgraded to 21 GW of capacity by 2015 (Global Times 2012). In addition, it is looking to marine renewables. China's tidal resource has been put at 190 GW, 38.5 GW of which is available for development, giving an annual output of 87 TWh of electricity. The China Ocean Energy Resources Division says 424 tidal power stations could be built along the coastline, mainly in maritime provinces like Zhejiang and Fujian (Guobing 2011).

All the above are focused on electricity, but China also has large solar thermal and biomass potential, some of which can be, and already is, harnessed at the local level, meeting heat needs directly, with over 100 GW(th) in place. Biogas production from agricultural and other wastes, including large pig farms, supplied 34 TWh in 2011. The great advantage of biogas is that it can be stored, so it is not a weather-dependent resource.

By contrast, as noted above, most of China's large green electricity assets are weather dependent and are also in remote locations, which is why supergrid links are needed. That can not only bring power to demand centres, but can also help with balancing the variable renewable inputs and variable demand, making use of the sheer size of the country. If it is not windy in one area it may well be in another.

The supergrid network may also be extended to provide links with other countries for exports and balancing imports when needed. Liu Zhenya, General Manager of the State Grid Corporation of China, has indicated that China plans to enhance transnational grid connection projects during 2011–15, and one of the key projects is a China–Russia direct-current scheme. China will, he said, accelerate construction of direct-current power transmission from Russia and Mongolia to Liaoning, Tianjin and Shandong during the periods 2016–20 and 2020–2030, based upon power transmission progress from surrounding countries to China and in accordance with development needs of north east China and north China (SinoCast 2012).

With up to 140 GW of hydro, and maybe even more wind capacity likely to be in place within the next 10 years, China may at times have some energy to spare, but also at times may need some balancing inputs. For that, there is abundant wind available in Inner **Mongolia**, and Mongolia has plans for developing large-scale CSP projects in the Gobi Desert. Excess electricity (around 1 GW) from the proposed Gobitec CSP project would be exported to urban centers in China, Japan, and South Korea via a new network of nearly 4000 km of HVDC transmission lines (Gobitec 2013). We could thus see the creation of an East Asian supergrid network, something that Japan in particular, being a series of islands with limited land area for renewable energy projects, may find very helpful.

Elsewhere in Asia the picture is more mixed. **South Korea** is still pressing ahead with nuclear power on a significant scale, aiming to expand from 40% to 56% over the next decade, and it is continuing to export its nuclear technology e.g. to **Vietnam**. However, it is also keen on renewables. For example it could have up to 2.5 GW of offshore wind capacity by 2019, according to data from the country's Ministry of Knowledge.

India's total 'new' renewables capacity was around 25 GW in 2010, plus around 37 GW of conventional large and small hydro, but it is also pushing nuclear hard, despite much local opposition. It is also planning to expand renewables, with about 900 MW of hydro projects in development, and by 2012 it had installed 18.4 GW of wind capacity. In April 2013, a new renewables target of 55 GW by 2017 was set, including solar, wind and biomass. In addition to solar heating/cooking, biomass is a widely used traditional source for heating, and biogas use has been extensively developed. In terms of power generation it is claimed that wastes and crop residues might support 16 GW (Abdulla 2011).

In 2009 the Indian government approved the Jawaharlal Nehru National Solar Mission, which aims to enable 20 GW of solar PV energy to be deployed by 2022. There is also interest in tidal power, for example in the coastal Indian state of Gujarat, which has green-lighted an initial 50 MW tidal power project (Renew India Campaign 2012).

Like India, **Pakistan** has a nuclear programme as well as some wind projects. Solar is widely used. Its renewable potential is vast, especially wind, perhaps 50 GW (TBL 2013).

After Fukushima, **Malaysia**, **Taiwan** and the **Philippines** decided to abandon their nuclear plans and, like many countries in the region, are exploiting their abundant renewable resources, most obviously solar. **Australia** and **New Zealand** have continued with their non-nuclear power policies. Both have very large renewable energy potentials: New Zealand with its existing hydro and wind and also possibly tidal current options, and Australia with its vast sun-drenched interior. Indeed, despite its relatively poor showing so far (it now has a 10%-by-2020 renewable electricity target) one optimistic study suggested that Australia could even get to zero carbon by 2020, using CSP and wind (ZCA 2010). Be that as it may, a government report has claimed that wind and PV would be the cheapest energy options by 2050, although it saw CSP as expensive (BREE 2012), and a range of 100% renewable scenarios have been explored and found to be technically viable (Elliston *et al* 2012). Meanwhile New Zealand wants to become the world's first carbon-neutral country. It plans to generate 90% of its electricity from renewables by 2025. They supply 65–70% at present (MFE 2013).

6.4 The Americas

The boom in shale gas extraction may dominate the news headlines, but renewable energy is also moving ahead rapidly in the **USA**. Renewables supply about 15% of US electricity, if off-grid use is included, and the potential for expansion is very large.

A report from the US National Renewable Energy Laboratory (NREL), found that the US renewable resource base was sufficient to support 80% renewable electricity generation by 2050, even in a higher-demand growth scenario. It also looked at a 90% option, with 700 GW of wind and solar PV. To accommodate this large variable supply input, the NREL said there would have to be major upgrades to the grid and up to 100 GW of balancing backup/load shifting/storage. But NREL's hourly modelling found that, with this backup in place, demand could always be met, even at peak times, although 8–10% of wind, solar and hydro generation would need to be curtailed, e.g. at times of low demand, under an 80%-by-2050 renewable energy scenario, and more storage would be needed in the 90% scenario. NREL said 'The direct incremental cost associated with high renewable generation is comparable to published cost estimates of other clean energy scenarios. Improvement in the cost and performance of renewable technologies is the most impactful lever for reducing this incremental cost' (NREL 2012).

As yet, unlike in Germany with its 80%-by-2050 target, there are no plans for expansion on anything like this scale in the US, although there are some independent proposals along similar lines (RMI 2012). Renewables are nevertheless moving ahead, in part since some are becoming commercially viable. Public support for renewables is also high, compared with nuclear, according to the poll by Pew, which found 49% against, 44% for nuclear in 2012, but 69% favouring more government support for renewables (AEI 2013).

Although the recession took its toll, with some company collapses, PV is approaching grid parity and is now expanding rapidly, with an annual market put at around 1 GW, while hydro remains a major source. However, wind power looks like being the main new renewable source, and on land wind is certainly moving ahead very rapidly in the US. By the end of 2012 there was around 60 GW in place, much of it in the mid-West and High Plains area, with very favourable cost trends (Wiser and Bolinger 2012). After a slow start, offshore wind could soon be catching up. The US government has announced $180m funding over four years, for four offshore wind projects, to accelerate the country's efforts in the sector, and aiming to cut costs. This is on top of the government plan, announced earlier, to invest $43m in improving technology and infrastructure in the offshore windpower sector. It has been claimed that the US could have 10 GW of offshore wind capacity by 2020, with 5 GW already planned and an offshore undersea HVDC grid proposed to link up east coast projects (Wind connection 2013).

Biomass and geothermal also continue to get support, and 2011 saw $3.4bn of Department of Energy (DOE) loan guarantees for four different CSP projects, with 1.8 GW in construction and 10 GW in the pipeline. There is also growing interest in wave and tidal power, with a DOE study suggesting that the US could ultimately get over 25% of its electricity from these sources (EPRI 2011). The US Water Power Program mission has plans for 23–38 GW output by 2039, supported by the Production Tax Credit (PTC) system and the Marine and Hydrokinetic Renewable Energy R&D Act. Consultants Pike Research see an upcoming marine power boom resulting in a total revenue of at least $161.2 million in North America.

By contrast the US nuclear programme seems to be going slowly. Under President Obama, the US government has supported expansion, in part by offering loan guarantees to prospective private developers. However, so far these have not been too successful (one company pulled out in 2010) and, following Fukushima, some projects were abandoned or delayed. Nevertheless, some still may go ahead, with work on two projects starting in 2013.

There are no specific targets for US nuclear expansion, other than the general aim to get 80% of energy from 'clean' sources by 2035, although in its provisional *Annual Energy Outlook 2011*, the DOE projected an increase in nuclear capacity of about 10 GW by 2035, about 10%, of which 6.3 GW would be new capacity (five reactors), the rest coming from up-rating. But given projected rises in demand and in other supply options, the overall nuclear share in electricity supply would fall from 20% to 17%.

Even if the projected expansion of nuclear does materialize, the path ahead seems clear. With their costs falling, renewables will lead. This has already happened. In 2011, a key milestone was passed, with US renewable electricity production being 18% more than that from nuclear, led by biomass and biofuels (46% of total renewables), followed by hydro (37%), and wind (13.4%). Since then it has moved further ahead (EIA 2013).

Canada is in a somewhat similar situation. It has a very large potential renewable resource, with around 60% of its electricity being supplied by hydro, but it also has a significant nuclear programme, based on its own CANDU technology, although some projects have recently been halted. In addition to hydro, wind and biomass, Canada has good tidal resources and has the world's first FiT for tidal projects. Greenpeace has claimed that renewables could rise from supplying around 17% of energy as now to 31% in 2020 and 71% in 2050, with over 80% of electricity coming from renewables by 2020 and 90% by 2050 (Greenpeace 2009).

In **South America** the picture is also mixed. Nuclear is widely and strongly opposed by the public. In 2011, 72% opposed it in Argentina, 79% in Brazil and 82% in Mexico. These are countries that already have nuclear plants, but opposition was also strong in the other countries in the region (BBC 2011), some of whom also had nuclear aspirations. Interestingly, following Fukushima, Brazil delayed its new nuclear plant plan.

Renewables are well established in the region, large hydro and biomass in particular. For example, **Brazil**, the region's leading economy, already gets 87% of its electricity from renewables, mostly hydro. It has 2.5 GW of wind capacity, with prices now said to be lower than for natural gas. The wind potential is estimated at 143 GW. PV, an obvious option, is only just starting out. Brazil produces ethanol fuel from sugarcane, some of it for export. But subsidised US corn-based ethanol is less expensive, so Brazil has been losing out. The biofuels versus food issue has also been influential (Timilsina 2012).

Mexico already has 2 GW of wind capacity and is looking to 12 GW by 2020, while a new 50 MW PV plant is claimed to generate at lower cost than coal. There are also some interesting new wave and tidal projects. Mexico aims to get 35% of its electricity from non-fossil sources by 2026, up from 20% now. In **Chile**, an election campaign promise by President Piñera was to get 20% of energy demand met from renewables by 2020. 82% of oil-rich **Venezuela's** electricity already comes from renewables, mainly hydro, but there are plans for expansion of wind power, with over 10 GW being said to be possible, and a 'sowing light' PV programme. **Peru** gets 56% of its electricity from hydro and is trying to build up wind, solar and biomass contributions via a FiT system. Nearly 100% of **Paraguay's** electricity comes from hydro, but it is trying to expand other renewables, as are **Patagonia**, **Bolivia** and **Ecuador**, with PV especially favoured.

Argentina, which gets 40% of its electricity from hydro, is building a 1350 MW wind farm with Chinese turbines, and is also pushing ahead with PV solar. **Colombia**, which currently gets 70% of its electricity from hydro, is investing in wind power. It has an estimated theoretical wind power potential of 21 GW. **Uruguay** plans to produce 90% of its electricity from renewables by 2015, 30% from wind, 45% hydro, and biomass 15%.

Moving north, **Nicaragua** aims to be 94% renewables based by 2017, using hydro and some wind. In the **Dominican Republic**, a 2007 law established tax breaks for investment in renewables, which account for 14% of electrical generation. **Cuba** has been slower off the mark, but there are many local off-grid PV projects, as well as hydro, wind and biomass resources, while solar and wind are obvious areas for development in the Caribbean region generally, with some projects underway or planned (CREDP 2013).

While funding remains an issue in much of Central America and the Caribbean, the solar and wind resources are large, as they are across the whole South American region. Large hydro dominates but the use of wind is spreading, as is PV, although more slowly (GTM 2013b), while the potential for using biomass, although controversial, is very large. Overall, the prospects for renewables in the region look very good (REW 2013).

6.5 Africa

Africa has amongst the world's best renewables potential, given its climate. As I have indicated earlier, solar, an obvious choice, is being taken up strongly in North Africa. Although **Egypt** has nuclear aspirations, as have some other North African states, the turmoil of the so-called Arab Awakening has put that further back on the agenda. By contrast, renewables have moved centre stage. Egypt aims to get 20% of its electricity from renewables by 2020, **Algeria** aims to have 22 GW of renewable capacity by 2030, and **Morocco** aims to get 40% of its electricity from renewables by 2020 (GTM 2013a). In addition to CSP and PV, there are also some wind projects, e.g. in Egypt, and there is even a novel pilot wind-to-gas hydrogen electrolysis project in Morocco.

By comparison, the story in sub-Saharan Africa is more mixed. **South Africa** gets around 6% of its electricity from nuclear, and wants to expand its capacity, although the recession led it to abandon the programme for the moment. Instead, renewables are being promoted, with 3.75 GW planned by 2016, mainly from biomass, wind, solar and small hydro. But much more is seen as possible (Banks and Schäffler 2006, Earthlife 2008). For example CSP is being looked at and it is suggested that there could be up to 10 GW of wave energy potential.

In the rest of sub-Saharan Africa, progress is even more mixed. In **Kenya**, the government seems keen to focus very heavily on nuclear, but is also backing renewables. It has large wind, solar and biomass resources, and is planning a 100 MW wave plant. It already has over 200 MW of geothermal capacity in place, and aims to meet 50% of its electricity needs with geothermal by 2018. **Nigeria**'s 2006 Renewable Energy Master plan has renewables supplying 13% of electricity in the short term, and 36% long term. **Niger** aims to get 10% of its primary energy from renewables by 2020 and **Senegal** 15% by 2025 (REN 21 2012).

In terms of technology, large hydro is dominant in many African countries (near 100% in some), but in addition to wind, micro-hydro is seen as attractive in some locations, while village-level PV projects have spread widely, for example in **Uganda**, **Tanzania**, **Chad**, **Rwanda**, **Angola, Gambia** and the **Congo**. **Ghana** has introduced a FiT for PV, and its 155 MW PV plant, opened in 2012, is the largest so far in Africa.

Renewable energy FiTs (REFiTs) are clearly helping roll-out renewable technology across Africa, as they have in the EU, but a recent non-govermental organisation (NGO) report argued that, to meet Africa's needs at the speed and scale required without burdening the energy poor, costs must be distributed across the population fairly, based on usage and ability to pay, while the international community can provide extra financial support, such as through 'top-up' payments via a Global REFiT Fund, in line with obligations under the UNFCCC/Kyoto protocol for repayment of climate debts (FoE 2012). There is also the Kyoto Clean Development Mechanism (CDM). Although some critics say that has not in practice supported many renewable energy projects, the UN says the CDM has delivered up to $43bn of foreign investment into developing nations since it started operating in 2006 (UNFCCC 2012). However, FiTs at national level seem more effective, with a report from the UN Development Programme seeing them as a key way to help reduce the financing costs of renewables (UNDP 2013).

Box 6.1. RECP (RECP 2010) plan for Africa

The Road Map for the Implementation of the EU–Africa Energy Partnership included three priority areas and the following targets:

Energy access: Africa and the EU will take joint action to bring access to modern and sustainable energy services to at least an additional 100 million Africans by 2020. This will be a contribution to the African objective of giving access to modern and sustainable energy to an additional 250 million people.

Energy security: Africa and the EU will take joint action to improve energy security by doubling the capacity of cross-border electricity interconnections and by doubling the use of natural gas in Africa, as well as doubling African gas exports to Europe.

Renewable energy and energy efficiency: Africa and the EU will take joint action such as:

- Building 10 000 MW of new hydropower facilities
- Building at least 5000 MW of wind power
- Building 500 MW of solar energy
- Tripling the capacity of other renewables
- Raising energy efficiency in Africa in all sectors

In terms of overall strategy and coordination, there has been no shortage of high-level initiatives on renewables in the region. In 2009, the EU–Africa Energy Partnership and the EU, together with the African Union, launched a 10-year Renewable Energy Cooperation Programme (RECP), see box 6.1, and announced a planned contribution of €5 million to start the programme (RECP 2010).

Backing the RECP, EU Commissioner for Development, Andris Piebalgs, said: 'We need a reliable source of electricity to fuel development. Africa has a vast untapped renewable energy potential, ranging from hydro, to solar, wind, geothermal and biomass which could be used to ensure millions of people have access to electricity'.

While the RECP plan is good news, 500 MW of PV solar is embarrassingly small, and 5 GW is not much of a 2020 target for wind. For example it has been claimed that Kenya alone could have 800 MW of wind generating capacity within three years and 3 GW ultimately. I have also seen wind resource estimates of 2.8 GW for Ghana, while South Africa could have much more. An Earthlife analysis suggested that South Africa could obtain over 50% of its power from renewables (400 TWh p.a.) by 2050, with the potential for on land wind being put at 50 GW, the wave potential at 10 GW, and the solar potential being enough in theory to supply the needs of the whole country (Holm 2008).

It is hoped things will change, with the RECP and subsequent programmes like the UN's new *Sustainable Energy for All* initiative, which includes €50m EU backing (UN 2013). IRENA is also offering support. But Africa needs more than top-down aid programmes. It needs local involvement, training and skill development to support the growth of local jobs, and technical and economic capacities; see box 6.2.

While renewables (large hydro apart) are still relatively marginal in Africa, there are signs of progress and good prospects for unlocking the huge resource (Hankins 2012,

Box 6.2. Africa's energy potential

'The solar radiation Africa receives could make this continent the Saudi Arabia of the future.' That was a conclusion from a 'Power Kick for Africa' strategy workshop on renewable energy policies organised by the World Future Council Foundation, in co-operation with the Energy Commission of Ghana. That brought together representatives from utilities, regulators, industry and civil society from ten African countries. They said they were determined to expand their co-operation under the umbrella of the African Renewable Energy Alliance (AREA 2013). Given that national grids are often unreliable or undeveloped in much of Africa, a key option is the development of local mini-grids to network locally available renewables (GVEP 2011).

IRENA 2013a). Nevertheless there are many problems to overcome, few of them technical. Institutional and political conflicts have bedevilled economic and social progress, as has the sheer size and social complexity of the continent (Onyeji *et al* 2012).

So far most new investment in energy systems in Africa seems to have been focused on large capital projects of uncertain social and environmental impact, the most familiar perhaps (leaving aside oil!) being the giant hydro project planned for the Congo river. And of course there is the continued push, often led by foreign vendors, for nuclear power. In addition to Kenya, at various points Ghana, Namibia, Nigeria, Senegal and Tanzania have all expressed an interest.

Does Africa need nuclear? At the RECP launch, EU Commissioner for Development, Andris Piebalgs noted that '1.6 billion people worldwide have no access to electricity, most of them in sub-Saharan Africa and southern Asia. Poor energy systems undermine growth potential in these countries from 1 to 2%'. With many people off the grid, very capital-intense centralised nuclear, supplied by foreign vendors, hardly seems to be the answer. Africa has abundant solar and other renewables, many of which can be developed effectively on a local decentralised basis. Their use offers an economically effective, socially equitable and environmentally sound way to meet local energy needs, cut emissions and aid social and economic development (McDonald 2008).

Globally, as I have described, it seems clear that, in most but not all cases, countries which are backing nuclear back renewables less, and vice versa. The obvious major exceptions are China and India although, in the former, the nuclear programme is relatively small compared to the renewables programme, while in the latter the nuclear programme has been slow-moving. Within the western EU, with France now perhaps changing sides to some extent, the UK, and to a lesser extent Finland, stand out as still being very pro-nuclear, but that position may not be sustainable. For example, while there are clear pressures to continue down the nuclear route in the UK, with proposals and scenarios for even more plants (ERP 2012), when German power company E.ON pulled out of the UK's Horizon nuclear project, it said 'We have come to the conclusion that investments in renewable energies, decentralised generation and energy efficiency are more attractive—both for us and for our British customers' (Teyssen 2012).

The writing seems to be on the wall. The economics of new nuclear are not looking good, at least in the EU (FT 2012). The situation elsewhere may be different, and governments in some countries, notably Russia, seem dedicated to a nuclear future. But everywhere renewables are getting cheaper and that process is unlikely to stop.

Some continue to argue that, if climate change is the key issue, then nuclear power is essential, as a low-carbon option. However, while the nuclear fuel cycle generates on average around 15 times less CO_2 than coal plants, it generates about 7 times more than modern wind turbines (Sovacool 2008). Moreover, this imbalance will get worse as reserves of high-grade uranium dwindle (it will take more energy to process lower-grade ores) and renewables become more energy efficient (Harvey 2010). A nuclear programme could thus see diminishing returns, while undermining renewables.

6.6 What next: the issues of cost and scale

As the global review above should have indicated, renewables are progressing quite well in many parts of the world, in the EU especially. Consultants Frost and Sullivan noted that, within the EU, by 2020 wind was expected to generate 647 TWh (up from 119 TWh in 2010), hydro 392 TWh (up from 327 TWh) and others 408 TWh (up from 124 TWh).

Meanwhile the output from EU nuclear plants was expected to fall from 937 TWh to 910 TWh between 2010 and 2020. So nuclear's share of total EU generation will fall from 28.0% to 23.7%. In addition, the use of coal for electricity generation would also fall, and quite dramatically, from 940 TWh in 2010 to 517 TWh by 2020 (Frost and Sullivan 2012). This sounds like good progress, unless you are a nuclear enthusiast, even if the use of fossil gas may yet expand in some countries, with shale gas playing a role. While the rise in fossil fuel use in some areas is worrying, generally renewables seem well placed.

Some still worry about grid balancing, particularly in the US, where distances are much larger and grids are much less integrated than in most of the EU. However a University of Delaware and Delaware Technical Community College study claimed that a well-designed combination of wind, solar power and storage in batteries and fuel cells would nearly always exceed electricity demands while keeping costs low. A computer model tested over four years of historical hourly weather data and electricity demands, using data from the large regional PJM Interconnection grid, which includes 13 states from New Jersey to Illinois, 20% of the US total electric grid. The model focused on minimizing costs instead of the traditional approach of just matching generation to electricity use (Budischak *et al* 2012).

It was found that generating more electricity than needed during average hours, in order to meet needs at high-demand but low-wind power hours, would be cheaper than storing excess power at high-wind periods for later high demand. The researchers claimed that 'using hydrogen for storage, we can run an electric system that today would meet a need of 72 GW, 99.9% of the time, using 17 GW of solar, 68 GW of offshore wind, and 115 GW of inland wind'. Moreover they found that aiming for 90% renewables by 2030 would cut net overall costs.

That is a bold claim, but similar views on the long-term, and also short-term, costs of renewables have emerged from other sources. In Germany, there has been much attention paid to what is sometimes called the 'merit-order effect', i.e. the tendency, given the FiT system, for the output from renewables like wind to displace output from fossil sources, thus reducing costs to consumers. It is claimed that the overall savings are greater than the FiT subsidy they pay for wind (Sensfuß 2007). The basic argument is that, if an investment portfolio approach is adopted, renewables like wind, with zero fuel costs, may win out (Awerbuch and Berger 2003).

A more recent comparative study by financial group Ernst & Young (E&Y) seems to confirm this. By factoring in returns to GDP, like jobs and local taxes, E&Y's analysis challenged the power sector's standard 'levelised cost of energy' (LCOE) approach, which you may recall I looked at briefly at the end of chapter 5 (see tables 5.1 and 5.2 for example). E&Y claimed that the *net* cost of European wind power was up to 50% lower than that of its main conventional power rival, combined cycle gas-fired plants. They noted that in Spain, producing 1 MWh will generate €56 of gross added value from wind, as opposed to €16 from combined cycle gas turbine (CCGT). Across the six EU focus countries (Spain, UK, France, Germany, Portugal and Poland), wind's net cost was competitive and, extrapolated across the EU as whole, actually cheaper (Ernst & Young 2012).

Following a similar *net cost* approach, a major renewables company, Mainstream Renewables, have pushed for recognition of the wider strategic benefits of offshore wind, for example in terms of security of supply and employment creation. They noted that a 2012 UK study of the 'Value of Offshore Wind' had looked at the benefits of investment in offshore wind. The study had found that, by 2015, it could increase UK GDP by 0.2%. It would also create over 45 000 full-time jobs. By 2020, it could increase GDP by 0.4%, and the number of people employed to over 97 000; and by 2030, add 0.6% to GDP growth, create 173 000 jobs, and deliver an increase in net exports of £18.8bn, covering nearly 75% of the UK's current balance of trade deficit (Mainstream 2012).

The conclusions from these assessments on the relative net cost of wind and its rivals may sound surprising although, for example, gas can be expensive in countries that have to import it. But even in the gas-producing UK, E&Y put wind's net cost only slightly above gas, at €35/MWh against €31/MWh. Moreover, that imbalance will reverse over time as gas resources, new and old, dwindle and wind technology improves.

Gas-fired CCGT plants do generate about 40% less CO_2/kWh than coal plants, so some see gas as a reasonable partner for renewables, while the latter develop, and the short-term availability of shale gas in some countries may add to that view. Even so, using gas still means more emissions and social and environmental costs, or the large economic cost of adding CCS systems. The data in table 5.2 in chapter 5 suggested that when CCS is added to gas plants, wind is clearly cheaper. It is also worth pointing out that these extra cost are uncertain, since no CCS plants of any size yet exist. It could cost more. Moreover it is not just the extra equipment at the gas plant that will add cost. CO_2 Europipe, an international consortium of companies and research institutions, claims that a network of 22 000 km of CO_2 pipelines would be needed across Europe, costing €50 billion. If these infrastructure costs are included, gas does not look like a cheap option (Beckman 2011).

It is true that renewables will also have infrastructure costs. For example, existing grid links will have to be strengthened and new supergrid links built. But as I argued earlier, these developments could pay for themselves by enabling export of excess electricity. By contrast CCS and its associated infrastructure is just a cost burden, making the continued use of fossil fuel look even more unattractive.

There may be no choice but to a try to deploy CCS in rapidly industrialising countries such as China, where coal use is likely to dominate for some time. Within the EU and US, some say that renewables and gas could complement each other for a while (RMI 2012). As noted in chapter 3, some environmentalists argue that, at some point in the future, if the technology is developed successfully, then CCS might be used with biomass-fired plants for negative net carbon operation. So fossil fuel CCS might be a useful stepping stone. However that is some way off and, although fossil gas will be needed in the short term to balance wind, longer term there are many other options. So most environmentalists are nervous about the expansion of gas. They would prefer to simply push ahead with renewables and non-fossil backup, so as to make the expanded use of gas, and of course coal, less vital, although that may not be possible in every area in the short term. It is a key current issue (Citi 2012, Elliott 2013b).

The bottom line on this issue, and indeed for the sustainable energy approach generally, is whether renewables can be expanded fast enough to meet needs. As this book should have indicated, the resource is there, the technology is ready or developing rapidly, and its costs are falling, with wind and PV now being competitive in some areas (Citi 2012). The experience with wind and PV in Germany, and now wind in China, shows that very rapid deployment is possible. Globally, wind capacity has grown at 20–30% per annum over several decades, PV is expanding even faster, albeit from a lower initial level. So, overall, it should be possible to continue to expand renewables rapidly on a large scale. To put that claim in perspective, the UN-linked International Sustainable Energy Organisation has calculated that, based on simple arithmetic, even assuming 2% per annum growth in energy use globally, renewables would only have to grow by an average of 5.2% per annum to be able to replace the use of all 'finite' fuels (fossil and nuclear) by 2050 (ISEO 2013).

On this view, what happens beyond 2050? There is a natural limit to renewable energy availability, defined (tidal and geothermal apart) by the amount of solar energy received by the planet. However, even assuming continued growth in demand, the total renewables resource limit is some way off, and many times current global energy use. There will be economic and environmental limits to what renewables can provide in practice, and there may well also be social, environmental and resource constraints on economic growth. However, assuming that it was sustainable and desirable, continued economic growth into the far future does not seem to be likely to be limited by basic energy scarcity, given the huge renewable energy resource.

Desertec have indicated that, in theory, all the world's current electricity use could be met from CSP plants covering the equivalent of an area of desert measuring about 338 km by 338 km square, and there are plenty of desert areas around the world. It notes that 'within six hours, deserts receive more energy from the sun than humankind uses in a year' (Desertec 2013). In addition, according to research by the Carnegie Institute and the Lawrence Livermore National Laboratory, in theory more than 400 TW of electricity could be extracted from surface winds and over 1800 TW from winds throughout the atmosphere (including via kites and other aerial devices). In total that is about 100 times more than total current global primary power demand (Marvel *et al* 2012).

These figures ignore load factors and other practical constraints, which would reduce the energy actually available. The only factors considered in the study were the geophysical limitations of the extraction techniques. There would also be social and environmental impacts from very large-scale use of solar or wind or, as I have indicated, biomass, including land use and potential biodiversity conflicts and possibly even effects on climate, due to the extraction of energy from the wind. The Carnegie/ Lawrence Livermore researchers mentioned above noted that if all humankind's electricity needs were met by wind turbines, that would change surface temperatures by around 0.1 degree Celsius and alter precipitation levels by around 1%.

Few see wind energy being used on this scale, and some suggest that, in practice, the total global resource may be much more limited (Adams and Keith 2013). However, taking into account the effects that turbines would have on surface temperatures, water vapour and other climatic impacts, a University of Delaware and Stanford University study found that about half of global electricity could be supplied from wind, with four million turbines of total capacity 5.75 TW, spread out worldwide, without significant eco-impacts (Archer and Jacobson 2012).

Bold visions like this might come up against other constraints, such as a scarcity of some of the key minerals and materials needed to construct the technologies. There could also be shortages of the energy needed to build them. That to some extent depends on how rapidly the expansion is carried out. Most materials can be recycled from earlier uses, given time, or in the case of some rare earth materials, substitutes used, or designs changed to avoid their use. Given that most of the energy for building renewable energy systems will initially have to come from fossil sources, some fear that the growth of renewables may falter when fossil fuels become scarce (Chefurka 2007). It is hoped, though, that sufficient renewables can be established to provide the energy to construct more capacity, in a self-sustaining 'breeding' process, so as to limit the use of fossil fuel, and their linked emissions. It should help that the Energy Return on Energy Invested (EROEI) ratios for renewables are mostly high (see section 4.3). But if low EROEI fossil fuels are to be burnt off, surely this energy should be invested in renewables as a priority?

The land-use issues may also lead to limits. By their nature, renewable energy flows are diffuse and the technology for capturing energy from the flows has to cover relatively large areas. It is instructive, and sobering, to revisit Professor David Mackay's calculations about the areas needed to match the energy needed per person from renewable sources (MacKay 2007). However this has to be put in perspective. Humankind has happily accepted large areas of land being used for farming, since we need food, and although we do have to avoid conflicts with that, there are many areas of marginal land not usable for cultivation, including deserts (for solar), as well as the sea (for wind, wave and tidal).

More prosaically, in the UK context, in response to claims that the countryside would be 'disappearing beneath solar panels', it was pointed out that the 200 acres of solar farms that had be established in the UK by 2011, were about 0.0003% of the UK land area, and that the 500 miniature 'crazy golf' courses in England, which typically covered roughly an acre each, accounted for twice the area covered by solar farms (Carrington 2011). If that calculation is expanded to include full-scale golf courses (over 0.5% of UK land area), that gives some idea of what the land use requirement might be for a major UK solar farm and wind programme, not that anyone is suggesting that golf courses be necessarily targeted. Remember too that the area around wind turbines can still be farmed (or used for golf!). Moreover, offshore wind and rooftop PV take up no land.

I looked at overall figures for PV and other renewables in chapter 4, but an amusing calculation by a UK PV trade lobbyist group suggested that only 1% of total UK land area would be required for enough solar arrays to meet the UK's entire electricity needs, though that ignores balancing and (night-time) backup requirements (Solar Portal 2012).

Biomass raises more significant land-use issues. To avoid the potentially unsustainable import of biomass, the Pugwash High Renewables Pathway envisaged 10% of UK land being used for biomass cultivation. Around 72% of UK land area is used for agricultural purposes (forestry excluded), and only some of that would be used for energy crops. However, there would have to be changes in farming practice and also perhaps diet (Pugwash 2013). That may be hard, but a less meat-based diet could be beneficial to health. As I reported in chapter 3, it has been suggested that, given proper regulation and the right choice of crop, somewhat similar patterns of changed land use might also emerge elsewhere in the world, without undermining food production, although vehicle biofuel options may, and perhaps should, be more constrained.

As I have indicated, there should be far fewer problems with solar. Above I focused on the UK. However, the same seems to be true in some other densely populated places. A study from the Indian Institute of Science Centre for Climate Change in Bangalore claims that 4.1% of the total uncultivatable and waste land area in India is enough to meet the projected annual electricity demand of 3400 TWh by 2070 using solar energy (Mitavachan and Srinivasan 2012). This assumed present-day PV technology. New solar cells will be more efficient and need less space per kWh produced. The study also did not take account of the fact that rooftop PV does not need any additional land or that some of the land around/under PV farm arrays can be used for other purposes like grazing. The authors added that the land area required would be further reduced to 3.1% 'if we bring the other potential renewable energy sources of India into the picture' e.g. wind and biomass. Overall then, it seems that, even in a highly populated and crowded country like India, renewables could in theory meet all electricity needs with minimal disruption to land use although, for biomass, clearly there will be biodiversity and water-use issues.

Finally, on land-use issues, I cannot resist relaying the view from US energy expert Amory Lovins that the land area/MW needed for nuclear plants, including their security enclosures, and their share of mining, fuel processing and waste disposal sites, is much larger than that for PV or wind projects, even leaving aside the very different embedded energy debts/EROEIs. He adds, provocatively, that 'a gram of silicon in amorphous solar cells, because they're so thin and durable, produces more lifetime electricity than a gram of uranium does in a light-water [nuclear] reactor' (Lovins 2009).

These views and visions are presumably not something that supporters of the conventional energy system would accept. Even the present level of commitment to renewables is often viewed with disdain. For many critics, renewables are a costly mistake and at best likely to make only a small contribution to meeting energy needs. For the foreseeable future, if there are environmental problems, then nuclear, gas and perhaps coal CCS, or even underground coal gasification (UCG 2013), can resolve them.

It was not my aim in this book to deal with the myriad issues surrounding conventional energy technology. However, as well as having looked at CCS briefly, I have dwelt at some length on the nuclear versus renewables issue, since it appears to be strategically vital to make choices about whether to back traditional large-scale centralised technology/fuels like nuclear, or the more decentralised renewable options. This issue has already been faced in Germany. Federal Minister of the Environment Norbert Röttgen said in 2010: 'It is economically nonsensical to pursue two strategies at the same time, for both a centralized and a decentralized energy supply system, since both strategies would involve enormous investment requirements. I am convinced that the investment in renewable energies is economically the more promising project. But we will have to make up our minds. We can't go down both paths at the same time'.

6.6.1 Issues of scale

The often tiresome nuclear versus renewables debate to some extent deflects us from the arguably much more important issue of future development paths and key strategic issues, such as which renewables to choose, how then to develop them and on what scale.

A long-running policy debate has been on whether to focus on large-scale or small-scale projects. Small and large are relative terms. Most current renewable energy projects are much smaller than conventional GW-sized conventional and nuclear projects. However some of the newly emerging offshore wind farms reach that scale, and some large hydro plants are already larger than that, and tidal barrages might also be large. This is a long way from the sort of small local community-based projects which many 'greens' have in mind.

Supergrid-based systems, especially if using energy imported from remote sources overseas, would presumably be viewed as even worse, as dangerously centralised and large scale. That was the view of celebrated German green energy pioneer Herman Scheer, who argued that Desertec-type projects would 'duplicate the current system' of centralised power and institutional control, whereby energy production and distribution was concentrated in the hands of a few multinational companies. He said 'we should be looking instead at decentralising the system, and looking closer to home for our energy supplies, such as solar panels on homes or harnessing wind energy on the coasts, or inland' (Scheer 2009).

The small-scale decentralist position has many merits. Technically, it can avoid long-distance transmission losses. Politically, it has become clear that locally controlled and owned projects are far more acceptable to local people than large projects imposed on them by corporate groups. That was one reason why the use of wind energy spread so quickly across Denmark. About 80% of the initial projects were locally owned by farmers or community co-ops. They recited the old Danish proverb 'your own pigs don't smell'. There were few objections. By contrast in the UK, where nearly all projects were and are owned by large companies, local opposition has been extensive and the funding system has made it very hard for small locally owned projects to get started.

In Germany, by the end of 2011, more than 65 GW of renewable electricity capacity had been installed, with more than half owned by citizens and farmers (WFC 2013). That included around half of the initial wind projects, and the advent of the FiT system has allowed many individuals to become direct owners of PV projects. In parallel, there has been a growth in local community-run, co-operatively owned renewable energy projects in villages and towns around the country. See box 6.3.

It is the same elsewhere, for example in Austria, where there are many local solar and biomass projects. As in Germany, there are many municipal projects. In a report on local

Box 6.3. Local energy projects in Germany

The number of energy co-ops in Germany has tripled to over 600 in two years, with more than 100 rural communities becoming 100% renewable energy based. Every second day a new one is being formed, according to Paul Hokenos (European Energy Review 2012). He reports that they now have some 80 000 active members who have invested at least €800m in green energy, mostly in solar, but increasingly in other renewables, such as wind power and hydro. The largest energy co-op, Stromrebellen in EW Schönau, has 130 000 electricity and 7000 gas customers.

There are many more, some of them using biomass/AD biogas as the main element, but some also using PV and wind (DGRV 2013). The village population of Feldheim near Berlin meet their electricity needs with wind, solar and biomass. The town of Bottrop in North-Rhine Westphalia is building a completely new infrastructure around renewables. Most of them are exporting energy, like Juehnde, one of several 'bioenergy villages', with 750 residents, which generates about 5 GWh pa, mainly via biogas, but only uses about 2 GWh. The largest so far are Rai-Breitenbach with 900 residents, Iden with 1000 and Randeqq with 1300 (Rai-Breitenbach 2013).

initiatives, the World Future Council noted that the Austrian town of Guessing now 'produces about 10 times more energy than it needs and approximately 40 times more electricity than it uses, all with renewable resources' (WFC 2013).

Despite the funding problems, with the advent of the small FiT, local co-ops have also begun to make an impact in the UK (Willis and Willis 2012) with, for example around 180 MW worth of projects in place or underway in Scotland, where grant aid has helped. The 2020 *Routemap for Renewable Energy in Scotland* included a target of 500 MW community and locally owned renewable energy projects by 2020. Overall it seems that, around the EU, 'local is good'. Clearly there are many local social and economic benefits from community-based projects, both in industrial countries, and also in developing countries (Practical Action 2013).

Nevertheless, not everyone is convinced that, technically, renewables can be developed sufficiently to meet all needs just on a small-scale local basis, using technologies like PV. A UK study suggested that community projects might meet up to 10% of total annual UK energy demands by 2020 (EST 2008). Certainly solar may well be suited to local small-scale deployment but, as I have pointed out, some of the other technologies are also much more efficient on a large scale, for example wind turbines. If they are located on higher wind speed sites than are available to most urban/suburban communities, their energy output will benefit from the square law on blade size and cube law on wind speed. Similar economies of scale apply to offshore wind, wave and tidal projects, most of which will be geographically remote from centres of population.

Even in the case of technologies that are technically viable on a small-scale individual or community basis, there can be issues of short-term costs. For example PV has been a major part of the decentralist vision, but it is still expensive. As a result there has been some strategic debate in Germany about whether PV should be allowed to expand as rapidly as it has so far. It is worth exploring this debate in more detail, since it illustrates some of strategic complexities that are opening up as the use of renewables expands.

6.6.2 The PV FiT debate

The German Advisory Council on the Environment (SRU) produced a detailed report setting out pathways for a transition to renewables. It concluded that it was possible to get to 100% renewable electricity by 2050, as against the government target of only 80% by then. However, the SRU claimed that the same result could be achieved if demand could be reduced from 700 TWh by 2050 to 500 TWh, with less PV then being needed. It was critical of how PV has been supported in Germany, arguing it was expanding too rapidly, imposing unsustainable costs. The current high rate of growth 'would result in half the capacity that is needed for a wholly renewable electricity supply in 2050 in the high electricity demand scenarios to be already installed in 2020. This means that unnecessary capacity would be installed prematurely, which in turn would increase long-term renewable electricity costs and jeopardise acceptance of a wholly renewable electricity supply'. Consequently, it said 'photovoltaic support should be drastically reduced so as to rectify mismanagement in this domain, whose current expansion rate far exceeds that deemed necessary' (SRU 2011).

Why the change of heart on PV? Germany has been at the forefront of PV deployment with capacity expanding rapidly as domestic-scale projects spread, stimulated by the FiT. That, the SRU said, was the problem. PV had boomed too fast, partly due to cost reductions, so that the high FiTs imposed too high a cost on consumers. This same argument has been heard across the EU, in Spain, Italy, France, and in the UK. They, like Germany, imposed FiT cut-backs or capacity caps.

Some saw this as just a failure of political nerve. They said the FITs should be left alone, since PV prices would then fall as the market grew, and the cost pass-through to consumers could also then be reduced. Certainly some might see throwing PV out of the mix as an odd idea. Economically it was almost certain to get very much cheaper.

In addition, the technical case sometimes made against PV is not that strong. PV is well matched to some energy demands. It can be used for lighting and computers in daytime occupancy buildings, for summer air-conditioning and even for topping up storage heaters for night-time use. More generally, although load factors are low, as more renewables with differing time signatures are linked into the grid, the balancing issue will become easier to handle. Moreover, solar is a huge resource, and the technology is well suited to rooftops, easy to install and run with no moving parts to go wrong. It may have been unwise to try to use FiTs to get its initial very high price down rapidly, but that does not mean the technology is at fault, or that FiTs are inappropriate, provided they are well designed, with effective price degression mechanisms.

Nevertheless, as I noted in chapter 4, with the worsening economic climate, the FiT for PV was seen by some as provocatively high and hard to defend. SRU said that further support in Germany was 'no longer justifiable on the grounds of learning curve effects, for the market for PV installations has grown considerably and is now international in scope. Even if Germany stopped promoting photovoltaic energy, the remaining PV installation market would be large enough to allow for further cost reductions'.

Although it had imposed high costs, the SRU was thus saying that the FiT had done its job and no more support was needed, or at least much less. The SRU did not suggest abandoning PV entirely. Instead it suggested that 'the scope of PV expansion should be kept at a low level that however still ensures that installed capacity can be adjusted to potential changes in demand. Only if a rise in electricity demand appears highly probable, PV capacity expansion should be promoted accordingly'. In effect, PV should be kept in reserve. It recommended that 'The PV development path should be mapped out in such a way as to allow for a timely response to trends in electricity demand'. Some might see the SRU's conclusion that 'PV support urgently needs to be reined in' as a capitulation to right-wing free-market enthusiasts or, more subtly, as a pragmatic response to likely political objections, which could rebound on other renewables. The SRU certainly hinted that 'perpetuating the current photovoltaic support framework would deprive renewables of funding that have the capacity to produce electricity far more efficiently'. Equally it might be seen as a sensible recognition of the limits of PV and FiTs in the current context. Most observers now agree that FiT levels for PV needed cutting, but the debate over how much continues.

I have gone into the SRU analysis of PV in some detail, not because I think it is necessarily right (PV does seem likely to play a major and expanding role in many places), but to highlight the need for a debate on how renewables expansion programmes are managed and supported, with pace and scheduling being key issues, i.e. which options should get priority. Given that there are many renewable energy programmes underway around the world, and PV is often seen as a key part, that debate is now becoming urgent.

There are many other urgent issues. It would for example be helpful to have more clarity on what are the best support systems to use. It is quite widely agreed that national level guaranteed price FiTs, as pioneered in Germany, have been very effective across the EU for getting wind capacity up fast at reasonably low cost, certainly more so than the UK's market-based Renewables Obligation trading system and probably more so than the USA's Renewables Portfolio Standard (RPS) quota target system (Mitchell *et al* 2006, Toke 2007, Szarka *et al* 2012).

If rapid deployment is required, in order to respond to the urgent climate change and energy security concerns, subsidies are needed in the short term for new technologies, to help them break into well-established markets dominated by the existing energy sources. But these are all meant to be interim subsidies. The real test of the schemes will be when they have achieved their capacity deployment and price reduction aims. Will it then prove easy to wean developers/users off them? In addition, are FiTs best suited to supporting the new, still high-cost, technologies? The evidence from PV is unclear, although now of course the situation for PV has changed since its costs have fallen. But what about wave and tidal current power? Would grant aid be best? Linked to successful energy output performance (Elliott 2012b)?

Moreover, what about wider multinational support schemes like the EU's Emission Trading System and the Kyoto Protocol-initiated Clean Development Mechanism (CDM)? The CDM may help some projects in developing countries but, in the absence of tight carbon caps, which have proved to be hard to negotiate globally, or even within the EU, carbon credit trading systems have not been too successful at cutting emissions or delivering renewables capacity, indicating the political limits and problems of global approaches. We may need new institutions and new initiatives, to speed up deployment.

There are many such issues that governments, agencies, researchers and developers will have to grapple with in the years ahead. Others include more technical questions such as whether emphasis should be on end-use efficiency or green energy supply, and the debate over the best way to balance variable renewables, via backup plants, energy storage, smart grids/supergrids or, rather, what mix of these balancing options should be used? I have covered some of these issues in earlier chapters, offering tentative conclusions. Some may be uncontentious. It is clear that smarter, stronger grids are needed to relieve congestion and help balance variable renewables (NREL 2009). That does have cost and impact implications, but also wider economic benefits, as with all infrastructure development. However, some issues are still very open, and the debate over aims, polices, priorities, strategies and tactics continues at various levels, from the reformist to the revolutionary (Abramsky 2010). I hope this book will lead readers to engage in it.

6.7 Conclusions

In reviewing the state of play with renewable energy, I hope I have managed to convey some of the excitement that exists in this rapidly expanding field. Whereas the global economic crisis is slowing activities in many areas, this is one sector that is growing rapidly, not least in terms of employment. By 2011 there were around 1.2 million people working in the sector in the EU (Eurobserver 2013). Globally, there are expected to be around 8 million people working just in the wind and solar sectors by 2020 (UNEP 2008).

Growth in employment in the renewables should more than replace the jobs lost by the phase-out of old energy systems. A 2009 study of the EU, looking at impacts on growth and employment, claimed that that the EU's target of supplying 20% of final energy consumption from renewables in 2020 could lead to a *net* rise of about 410 000 additional jobs and a 0.24% *net* rise in gross domestic product (EmployRES 2009). As I noted earlier, subsequent studies have put the economic benefits even higher, given savings on increasingly expensive imported fossil fuel. It is an obvious area for growth, as has been pointed out by environmental and trade union lobby groups (CaCC 2010).

Seeking rapid growth does, however, have its problems. Some technologies may not pan out and some may prove problematic. Moreover, growth in itself can be problematic. Some environmentalists look to a stable state future, with energy demand managed in a sustainable and equitable way and only ecologically sustainable growth allowed. To get there, renewable energy technologies will have to be deployed on an expanding scale for many decades, to replace existing systems. Many want faster progress (IRENA 2013b).

I have looked at some of the problems that this might entail, but no doubt many more will emerge (IEA 2013). I have looked at some of the strategic issues, such as the conflict with nuclear. However I have not gone into the battle over shale gas. This could be seen as the 'last gasp' of the fossil fuel regime and may divert attention from renewables for a while (Elliott 2011a). But it is not a long-term resource, and in the short term, using it will raise emissions, while fracking may have local environmental impacts (EC 2012).

I have not looked at thorium, which some see in a similar way as the nuclear industry's shale gas, as a last gasp. Thorium is not fissile so it has to be converted into a fissionable form, e.g. by mixing it with plutonium, and there are many unresolved issues. Although some are optimistic, for example about molten fluoride fast reactors (EFTF 2012, Weinberg 2013), others point to potential problems (Ashley 2012), and a commercially available and viable plant is some way off (Elliott 2011b, Tickell 2012).

I looked briefly at energy efficiency, but I have not dealt with the much broader issue of overall consumption and economic growth. Can and should it continue to grow forever? Technology can help to reduce impacts and waste and, up to a point, renewables offer a way to expand energy use, if that is what is wanted. But there may be other environmental and resource limits which technology cannot avoid. Social and lifestyle changes may also be needed. If so, how radical must they be, how soon must they be adopted, and do they include population control (Elliott 2003, Jackson 2009, OPT 2013)?

Although transport was covered in some of the scenarios I have mentioned (see box 5.1), the transport issue is a hard nut to crack; affluence and ease of travel are often seen as inseparable, and fuel use in this sector continues to rise inexorably, as do emissions. A modal shift away from cars to renewable energy-powered public transport seems vital but, in addition to cycling and walking, some technical solutions may also help in terms of personal/local mobility. There are proponents of both green gas and green electricity for powering personal vehicles (we may need both) and surely all trucks, vans and buses ought to run on biogas or green gas. However, unless some clever new low-impact biofuel/synfuel emerges, we may have to accept less global mobility, with less growth of cheap short-haul flights. That said, flying is perhaps unfairly singled out as a key issue. It is perverse that it is one of the few examples of an essentially untaxed fuel. Even so, the difference in full system emissions/passenger mile between plane and train can be small compared with the emissions from single occupancy cars (Chester and Horvath 2009).

There is also the more general issue of whether we should be focusing so much just on carbon reduction. What about other emissions, for example of methane from farming, fracking and even maybe undersea methane hydrate mining? And also of radioactive material from nuclear plants? And crucially, what about water resources? All steam-raising energy systems need cooling, and maintaining access to water will become a major problem for fossil and nuclear plants as climate change impacts (WRG 2009).

It already is. Hydro outputs are falling in many parts of the world. Several nuclear plants in France and the US have had to shut down during hot summers. One coastal US plant had to close down because the seawater it used for cooling had reached 74°F (23°C). Interestingly, plants with CCS use a lot of water, up to four times more per kWh than without CCS (Tzimas 2011). It is true that some renewables also need cooling water to work efficiently, CSP for example, and of course, stable rainwater supplies are vital for hydro plants (DoE 2006). But fortunately, most renewables do not need water for cooling and some can power seawater desalination. And none produces radiation.

Some governments and some lobby groups still seem unconvinced that renewables are viable on a significant scale, this view sometimes coinciding with support for nuclear power and/or scepticism about the significance or cause of climate change (GWPF 2013).

However, even if the climate change issue is ignored or downgraded, it can be argued that the use of renewables still helps to avoid the many other social and environmental problems associated with the use of conventional fossil and nuclear fuel and, crucially can replace these finite fuels as they become depleted. Moreover, it can be argued that, even leaving environmental and resource issues aside, longer term, renewables can win out economically. Indeed, in some cases and locations, they already represent the best, lowest cost option, regardless of climate issues, and despite the continued subsidies to fossil and nuclear-based energy technology (IRENA 2013c).

As I explore briefly in the final *afterword* chapter, there are many who doubt assertions like this. In this book I have tried to present a coherent and, I hope, convincing case that they are wrong. It has inevitably, in many cases, been a review of 'work in progress', and future plans. Nevertheless, overall, despite the problems and the unknowns, the bigger picture should now be a little clearer. As I have tried to show, so far, with often very minimal support, renewables have demonstrated that they can develop rapidly and that there is potential for even more rapid expansion around the world. So it could be argued that renewables, along with energy efficiency, should now be given a chance, and the necessary funding, to show what they can do.

Box 6.4. Do it yourself—some web links

I have described how energy systems are being changed, or may be changed, but you might like to get more direct access to see how it all works in practice and even try your hand at designing a better energy system.

First off, for a simple snapshot of how grid systems are balanced for small perturbations, with small changes in frequency, see the live UK grid frequency balancing website: http://www.dynamicdemand.co.uk/grid.htm.

A bit more fun, watch PV solar inputs over (day) time in Germany: http://www.sma.de/en/news-information/pv-electricity-produced-in-germany.html. There are also some fascinating dynamic charts showing daily power supply patterns in Germany over the past year (scroll down to see them) at http://www.ise.fraunhofer.de/en/downloads-englisch/pdf-files-englisch/news/electricity-production-from-solar-and-wind-in-germany-in-2012.pdf. Note it is a big file but worth opening.

And if you want to try your hand at building a UK 2050 energy scenario see http://www.decc.gov.uk/my2050. This is a much simplified version of DECCs full '2050 Pathways' model, which you can also tinker with by choosing different supply and demand levels. It includes access to a web tool and Excel calculator. Be warned: it is quite complex! http://webarchive.nationalarchives.gov.uk/20121217150421/ http://decc.gov.uk/en/content/cms/tackling/2050/2050.aspx.

You can store and share your results on the system and see previous efforts. The High Renewables version I worked on is in the 2013 Pugwash report, http://www.britishpugwash.org/recent_pubs.htm.

If you want to look more widely, IRENA's Global Atlas for wind and solar is useful—it allows you to check resources and local constraints around the world: http://www.irena.org/GlobalAtlas

Finally, if you need updates to the renewables story I have outlined in this book, see my weekly *Renew Your Energy* blog, http://environmentalresearchweb.org/blog/renew-your-energy/ and my free bimonthly newsletter *Renew On Line*: http://www.natta-renew.org.

In 1981, Professor Stephen Salter, a UK wave energy pioneer, said 'We are attempting to change a status quo which is buttressed by prodigious investment of money and power and professional reputations. For 100 years it has been easy to burn and pollute. 100 years of tradition cannot be swept away without a struggle. The nearer renewable energy technology gets to success, the harder that struggle becomes' (Salter 1981).

Much progress has been made since 1981, but sadly I think Salter's warning remains relevant and there will still be battles ahead. I hope, though, that this book will help to 'renew your energy' for the next phase, in which case box 6.4 may be of interest.

Summary points

- Renewables are expanding around much of the world, but often less so in countries that are also keen on nuclear: there can be conflicts.
- A major driver for their expansion is that many renewables are getting cheaper than conventional sources, inevitably perhaps since they do not incur fuel costs.
- The renewable resource is very large, although there may be ultimate constraints on how much we can and want to use.
- Many issues need to be addressed, concerning how the expansion of renewables is managed, e.g. which to focus on and how to provide the necessary support.
- There is also the issue of scale: local is good but (some) large may be more efficient.
- Energy efficiency and end-use issues are vital, as is sustainable consumption.
- Renewables are still fighting an uphill battle against the established, often heavily subsidised, energy technologies, but progress is being made.
- You can get involved!

References

Abramsky K (ed) 2010 *Sparking a World-Wide Energy Revolution* (Oakland: AK Press)

Abdulla F 2011 A renewable future for mankind: challenges and prospects *Making It Magazine*, Jan 11, http://www.makingitmagazine.net/?p=2849

Adams A and Keith D 2013 Are global wind power resource estimates overstated? *Environ. Res. Lett.* **8** 1

AEI 2013 Polls on Environment, Energy, Global Warming and nuclear power, American Enterprise Institute for Public Policy Research, http://www.scribd.com/doc/136515942/Polls-on-the-Environment-Energy-Global-Warming-and-Nuclear-Power-AEI-Public-Opinion-Study-April-2013#page=92

Archer C and Jacobson M 2012 Saturation wind power potential and its implications for wind energy *Proc. Natl. Acad. Sci.* **109** 15679–84, September 25

AREA 2013 African Renewable Energy Alliance, http://area-network.ning.com/?xg_source=msg_mes_network

Ashley S, Parks G, Nuttall W, Boxall C and Grimes R 2012 Nuclear energy: thorium fuel has risks, *Nature* **492** 31–3

Awerbuch S and Berger M 2003 Applying Portfolio Theory to EU Electricity Planning and Policy Making, OECD/IEA, http://www.awerbuch.com/shimonpages/shimondocs/iea-portfolio.pdf

Banks B and Schäffler J 2006 The potential contribution of renewable energy in South Africa, Report for Earthlife/SECCP, Feb

BBC 2011 Opposition to Nuclear Energy Grows: Global Poll, BBC World Service GlobeScan survey, http://www.globescan.com/news_archives/bbc2011_energy/

Beckman K 2011 Finally, the plan for the CCS revolution *European Energy Review*, Nov 17, http://www.europeanenergyreview.eu/site/pagina.php?id=3360

BREE 2012 Australian Energy Technology Assessment, Australian Government Bureau of Resources and Energy Economics, http://www.bree.gov.au/publications/aeta.html

Budischak C, Sewell D, Thomson, Mach L, Veron D and Kempton W 2012 Cost-minimized combinations of wind power, solar power and electrochemical storage, powering the grid up to 99.9% of the time *J. Power Sources* **225** 60–74

CaCC 2010 One Million Jobs Now, UK Campaign against Climate Change report, http://www.campaigncc.org/greenjobs#pamphlet

Carrington D 2011 The new mutations of renewable energy nimbys, Environment Blog *The Guardian*, Aug 30, http://www.guardian.co.uk/environment/damian-carrington-blog/2011/aug/30/nimby-solar-wind-energy

Chefurka P 2007 World Energy and Population, web paper, http://www.paulchefurka.ca/WEAP/WEAP.html

Chester M and Horvath A 2009 Environmental assessment of passenger transportation should include infrastructure and supply chains *Environ. Res. Lett.* **4** 024008

Citi 2012 Shale & renewables: a symbiotic relationship, Citi Research, London, https://ir.citi.com/586mD%2BJRxPXd2OOZC6jt0ZhijqcxXiPTw4Ha0Q9dAjUW0gFnCIUTTA%3D%3D

CREDP 2013 Caribbean Renewable Energy Development Programme, St Lucia, http://www.credp.org/

Danish Government 2011 Energy Strategy 2050, The Danish Government, http://www.ens.dk/Documents/Netboghandel%20-%20publikationer/2011/Energy_Strategy_2050.pdf

Danish Government 2012 Danish Climate and Energy Policy Report 2012, Report from the Ministry of Climate, Energy and Building to the Danish Parliament, http://www.ens.dk/en-US/policy/danish-climate-and-energy-policy/Documents/Energy%20Policy%20Report%202012.pdf

DECC 2012 Renewable Energy Roadmap, UK Department of Energy and Climate Change, https://www.gov.uk/government/publications/uk-renewable-energy-roadmap-update

DENA 2012 'Efficient integration of renewable energy into future energy systems', Deutsche Energie-Agentur GmbH - the German Energy Agency (DENA), http://www.dena.de/studien

Desertec 2013 Desertec Foundation, http://www.desertec.org

DGRV 2013 'Energy Co-operatives' DGRV-Deutscher Genossenschafts und Raiffeisenverband e.V., Berlin, http://xa.yimg.com/kq/groups/20593576/937211628/name/Energy_Cooperatives%20DGRV%202012%20how%20to%20Establish1%2Epdf

DoE 2006 Energy demands on water resources, Report to Congress on the interdependency of energy and water, US Department of Energy, http://www.sandia.gov/energy-water/docs/121-RptToCongress-EWwEIAcomments-FINAL.pdf

Earthlife 2008 Sustainable Energy Briefings, Earthlife, http://www.earthlife.org.za/

EC 2012 Environmental Aspects on Unconventional Fossil Fuels, European Commission reports, http://ec.europa.eu/environment/integration/energy/unconventional_en.htm

Edano Y 2012 Ministers comment in August, quoted in the Japan Times, http://www.japantimes.co.jp/text/nn20120823f1.html

EFTF 2012 US Energy from Thorium Foundation website, http://energyfromthorium.com

EIA 2013 US Energy Information Administration, monthly energy data, http://www.eia.gov/totalenergy/data/monthly

Elliott D 2003 *Energy, Society and Environment* (London: Routledge)

Elliott D 2010 *Nuclear or Not?* (Basingstoke: Palgrave Macmillan)

Elliott D 2011a Fracking Hell or Gas Delight *Renew Your Energy* blog, Environmental Research Web, May 21, http://environmentalresearchweb.org/blog/2011/05/fracking-hell-or-gas-delight.html

Elliott D 2011b Is thorium the answer? *Renew Your Energy* blog, Environmental Research Web, June 18, http://environmentalresearchweb.org/blog/2011/06/is-thorium-the-answer.html

Elliott D 2012a Fukushima: Impacts and Implications (Basingstoke: Palgrave Macmillan)

Elliott D 2012b 'Windpower: opportunities, limits and challenges' chapter in Szarka J *et al Learning from Windpower: Governance and Societal Perspectives on Sustainable Energy* (Basingstoke: Palgrave Macmillan)

Elliott D 2012c Greening Japan's Energy *Renew Your Energy* blog, Environmental Research Web, July 14, http://environmentalresearchweb.org/blog/2012/07/greening-japans-energy.html

Elliott 2013a Scotland – isolated or leading *Renew Your Energy* blog, Environmental Research Web, March 23, http://environmentalresearchweb.org/blog/2013/03/scotland-isolated-or-leading.html

Elliott 2013b Gas and air *Renew Your Energy* blog, Environmental Research Web, March 9, http://environmentalresearchweb.org/blog/2013/03/gas-and-air.html

Elliston B, Diesendorf M and McGill I 2012 Simulations of scenarios with 100% renewable electricity in the Australian National Electricity Market *Energy Policy* **45** 606–13, http://www.sciencedirect.com/science/article/pii/S0301421512002169

Employ RES 2009 EmployRES—The impact of renewable energy policy on economic growth and employment in the European Union, Report for the European Commission's Directorate-General Energy and Transport, produced by Fraunhofer ISI, Ecofys, Energy Economics Group, and others, http://ec.europa.eu/energy/renewables/studies/doc/renewables/2009_employ_res_report.pdf

EPRI 2011 Mapping and assessment of US Ocean wave energy resource, Electric Power Research Institute, http://www1.eere.energy.gov/water/pdfs/mappingandassessment.pdf

Ernst & Young 2012 Analysis of the value creation potential of wind energy Policies: A comparative study of the macro-economic benefits of wind and CCGT power generation, Ernst and Young consultants report for ACCIONA—EDP, Spain, May, Link from http://www.eurotrib.com/story/2012/11/1/65817/6658

ERP 2012 Nuclear Fission Technology Roadmap, Energy Research Partnership report, produced by the UK National Nuclear Labs at Sellafield 2012, http://www.energyresearchpartnership.org.uk/nucleartechnologyroadmap

EST 2008 Power in Numbers, Energy Saving Trust, London, http://www.energysavingtrust.org.uk/Publications2/Local-delivery/Legislation-and-policy/Power-in-numbers-summary-report

Ethics Commission 2011 Germany's Energy Turnaround: A collective effort for the future, Ethics Commission on a Safe Energy Supply, report to the German Federal government, May

Eurobserver 2012 Eurobserver EU statistics service, press release, http://www.eurobserv-er.org/pdf/press/year_2012/RES/English.pdf

Eurobserver 2013 The State of Renewable Energies in Europe, Eurobserver EU statistics service, http://www.eurobserv-er.org/pdf/bilan12.asp

Eurobarometer 2013 Attitudes of Europeans towards Air Quality, Flash Eurobarometer 360, European Commission, http://www.ec.europa.eu/public_opinion/flash/fl_360_en.pdf

Europa 2013 Renewable energy share of renewable energy up to 13% of energy consumption in the EU27 in 2011, Europa press release on Eurostat data, April 26, http://europa.eu/rapid/press-release_STAT-13-65_en.htm

European Energy Review 2012 Germany's Little Energy Co-ops Make a Big Splash, European Energy Review, Dec 17, http://www.europeanenergyreview.eu/site/pagina.php?id=4005&zoek

FoE 2012 Powering Africa through Feed In Tariffs, Friends of the Earth UK, WWF, http://www.foe.co.uk/resource/reports/powering_africa_summary.pdf

Fraunhofer 2012 Electricity production from solar and wind in Germany, Fraunhoffer Institute/ISE, http://www.ise.fraunhofer.de/en/downloads-englisch/pdf-files-englisch/news/electricity-production-from-solar-and-wind-in-germany-in-2012.pdf

Froggatt A, Mitchell C and Managi S 2012 Reset or Restart? The Impact of Fukushima on the Japanese and German Energy Sectors, Chatham House Briefing Paper, Chatham House, London, http://www.chathamhouse.org/publications/papers/view/185005

Frost and Sullivan 2012 European Nuclear Power Sector: Trends and Opportunities, Consultants Frost & Sullivan, Oct, http://www.frost.com

FT 2012 EDF plant cost rises damp nuclear hopes, Financial Times, London, Dec 3, http://www.ft.com/cms/s/0/662e0b40-3d68-11e2-9f35-00144feabdc0.html#axzz2JC7MlATL

GCRI 2012 Germany's Transition: One year later, German Centre for Research and Innovation, http://www.germaninnovation.org/news-and-events/past-events/past-event?id=c8b972a5-5f95-e111-9c57-000c29e5517f

GEA 2012 Global Energy Assessment, International Institute for Applied Systems Analysis, Austria, http://www.iiasa.ac.at/web/home/research/researchPrograms/Energy/Home-GEA.en.html

Global Times 2012 China's PV industry boosted by government support, Global Times, Dec 22, http://www.globaltimes.cn/content/751664.shtml

Gobitec 2013 Gobitec project website, http://www.gobitec.org/

Greenpeace 2009 Greenpeace and Renewable Energy Industry Call for Energy Revolution, Greenpeace Canada, http://www.greenpeace.org/canada/en/recent/call-for-energy-revolution/

Gruber A 2010 The costs of the French nuclear scale-up: A case of negative learning by doing *Energy Policy* **38** 5174–88, http://www.sciencedirect.com/science/article/pii/S0301421510003526

GTM 2013a Middle East and North Africa (MENA) Solar Market Outlook, 2013-2017, Greentech Media/Emirates Solar Industry Association, http://www.greentechmedia.com/research/report/mena-solar-market-outlook-2013-2017

GTM 2013b Solar in Latin America & The Caribbean 2013: Markets, Outlook & Competitive Positioning, Greentech Media, Boston, http://www.greentechmedia.com/research/report/solar-in-latin-america-the-caribbean-2013

Guobing Z 2011 China's tidal power development speeds up *China Daily*, Oct 25, http://english.peopledaily.com.cn/90778/7626191.html

GVEP 2011 Mini grid development, GVEP International Policy Briefing, http://www.gvepinternational.org/sites/default/files/policy_briefing_-_mini-grid_final.pdf

GWPF 2013 Global Warming Policy Foundation, London, http://www.thegwpf.org

Hankins M, Gustavsson M and Hinrichs F 2012 Unlocking Africa's Renewable Energy Potential, Renewable Energy World, Oct 2, http://www.renewableenergyworld.com/rea/news/article/2012/10/unlocking-africas-renewable-energy-potential?cmpid=WNL-Wednesday-October3-2012

Harris A 2013 How to solve the energy crisis *Eng. Tec. Mag.*, IET **8** (1) Jan 21, http://eandt.theiet.org/magazine/2013/01/how-to-energy.cfm

Harris G, Heptonstall P, Gross R and Handley D 2012 Cost estimates for nuclear power in the UK, ICEPT Working Paper WP/2012/014, Imperial College, London, August

Harvey D 2010 Carbon-Free Energy Supply (London: Earthscan)

HMG 2013 The UK's Nuclear Future HM Government, www.gov.uk/government/organisations/department-for-business-innovation-skills/series/nuclear-industrial-strategy

Holm D, Banks D, Schäffler J, Worthington R and Afrane-Okese Y 2008 Renewable Energy Briefing Paper: Potential of Renewable Energy to contribute to National Electricity Emergency Response and Sustainable Development, Earthlife, http://www.earthlife.org.za/

IEA 2013 Tracking Clean Energy Progress 2013, International Energy Agency, Paris, www.iea.org/publications/TCEP_web.pdf

IES 2010 Energy Strategy of Russia until 2030, Ministry of Energy of the Russian Federation, Institute of Energy Strategy, Moscow, http://www.energystrategy.ru/projects/docs/ES-2030_(Eng).pdf

IPSOS 2011 Global Citizen Reaction to the Fukushima Nuclear Plant Disaster, IPSOS Global Advisor, global poll carried out in May, published in June, http://www.ipsos-mori.com/Assets/Docs/Polls/ipsos-global-advisor-nuclear-power-june-2011.pdf

IRENA 2013a Africa's Renewable Future: The Path to Sustainable Growth, International Renewable Energy Agency, Abu Dhabi, http://www.irena.org/menu/index.aspx?mnu=Subcat&PriMenuID=36&CatID=141&SubcatID=276

IRENA 2013b Doubling the Global Share of Renewable Energy: A Roadmap to 2030, International Renewable Energy Agency, Abu Dhabi, http://irena.org/DocumentDownloads/Publications/IRENA%20REMAP%202030%20working%20paper.pdf

IRENA 2013c Renewable Power Generation Costs, International Renewable Energy Agency, Abu Dhabi, http://www.irena.org/menu/index.aspx?mnu=Subcat&PriMenuID=36&CatID=141&SubcatID=261

ISEO 2013 International Sustainable Energy Organisation, Geneva, http://www.uniseo.org/

Jackson T 2009 Prosperity without growth, Sustainable Development Commission report, London, http://www.sd-commission.org.uk/publications/downloads/pwg_summary_eng.pdf

Japan Times 2012 Japan to aim for zero nuclear power reliance in 2030s, Japan Times, Sept 13, http://info.japantimes.co.jp/text/nn20120913a6.html

JWPA 2010 Long-Term Installation Goal on Wind Power Generation and Roadmap V2.1, Japan Wind Power Association, http://jwpa.jp/page_132_englishsite/jwpa/detail_e.html

Lovins A 2009 Four Nuclear Myths, Rocky Mountain Institute, www.rmi.org/images/PDFs/Energy/2009-09_FourNuclearMyths.pdf

Ma H and Fu L 2011 Beyond the Numbers: A Closer Look at China's Wind Power Success, Worldwatch Institute, Washington DC, Feb 28, http://blogs.worldwatch.org/revolt/beyond-the-numbers-a-closer-look-at-china%E2%80%99s-wind-power-success/

MacKay D 2007 Sustainable energy without the hot air, free online version, http://www.withouthotair.com/

Mainstream 2012 Capturing the value of offshore wind, Mainstream renewables, http://www.mainstreamrp.com/content/reports/capturing-the-value-of-offshore-wind.pdf

Marvel K, Kravitz B and Caldeira K 2012 Geophysical limits to global wind power *Nature Climate Change*, on line Sept. 9, http://www.nature.com/nclimate/journal/vaop/ncurrent/full/nclimate1683.html

Maue G 2012 Presentation on Germany's energy programme to a PRASEG seminar in February, London, Parliamentary Renewables and Sustainable Energy Group

McDonald D (ed) 2008 Electric capitalism: recolonization in Africa on the power grid, http://www.hsrcpress.ac.za/product.php?productid=2243

Menasol 2013 PV guide to the MENA region, Menasol, Middle East and North Africa Solar Conference/PV insider, http://www.pv-insider.com/mena

MFE 2013 National Policy Statement for Renewable Electricity Generation, New Zealand Ministry for the Environment, http://www.mfe.govt.nz/rma/central/nps/generation.html

Mitavachan H and Srinivasan J 2012 Is land really a constraint for the utilization of solar energy in India? *Current Science*, **103** (02) July 25, http://www.currentscience.ac.in/Downloads/download_djvu.php?titleid=id_103_02_0163_0168_0

Mitchell C, Bauknecht D and Connor P 2006 Effectiveness through risk reduction: a comparison of the renewable obligation in England and Wales and the feed-in system in Germany *Energy Policy* **34** 297–305

Modern Power Systems 2011 Weak grid connections stalling China's wind energy growth *Modern Power Systems* Feb 27, http://www.modernpowersystems.com/story.asp?sectioncode=131&storyCode=2058965

NREL 2009 Wind Energy Curtailment Case Studies May 2008-May 2009, US National Renewable Energy Labs http://www.nrel.gov/docs/fy10osti/46716.pdf

NREL 2012 Renewable Electricity Futures Study, US National Renewable Energy Labs, Golden, http://www.nrel.gov/analysis/re_futures/

Onyeji I, Bazilian M, Nussbaumer P 2012 Contextualizing electricity access in sub-Saharan Africa *Energy for Sustainable Development*, **16** (4), December, 520–527, http://www.sciencedirect.com/science/article/pii/S0973082612000646

OPT 2013 Optimum Population Trust (now 'Population Matters') website, http://populationmatters.org/

Practical Action 2013 Intermediate Technology Development Group website, http://www.practicalaction.org/

Pugwash 2013 Pathways to 2050: Three possible UK energy Strategies, British Pugwash, London, http://www.britishpugwash.org/recent_pubs.htm

Qi W 2011 China plans $1.8 trillion wind power plan for 2050 *Windpower Monthly*, Oct 20, http://www.windpowermonthly.com/article/1099715/China-plans-18-trillion-wind-power-plan-2050

RECP 2010 Renewable Energy Cooperation Programme http://www.africa-eu-partnership.org/newsroom/all-news/energy-africa-Lauch-renewable-energy-cooperation-programme

Renew India Campaign 2012 Renewable Energy Installed Capacity in India Reaches 26,368.36 MW, Renew India Campaign, information portal, Kerala, http://www.renewindians.com/2012/12/Renewable-Energy-Installed-Capacity-in-India.html

REN21 2012 Global Status report, Renewable Energy Policy Network for the 21 Century, Paris, http://www.ren21.net/gsr. And also http://fs-unep-centre.org

Reuters 2012 Greenhouse gas emissions off 2.1 pct in 2011, http://www.reuters.com/article/2012/04/12/germany-emissions-idUSL6E8FC53420120412

REW 2013 Latin America Report *Renewable Energy World*, Feb 20, http://www.renewableenergyworld.com/rea/news/article/2013/02/latin-america-report-mexico-just-scratching-the-surface-of-solar-demand?cmpid=WNL-Friday-February22-2013

Rai-Breitenbach 2013 German Bioenergy village website, http://www.bioenergiedorf-odenwald.de/english/news/

RMI 2012 Reinventing Fire, Rocky Mountain Institute, Colorado, http://www.rmi.org/ReinventingFire

Salter S 1981 Wave energy: problems and solutions, *J. Royal Soc. Arts Proc.*, London, Aug 580

Scheer H 2009 Hermann Sheer as reported in *The Guardian* 17/6/09

Scottish Government 2011 Routemap for Renewable Energy in Scotland 2011, The Scottish Government, Edinburgh, June, http://www.scotland.gov.uk/Publications/2011/08/04110353/0

Scottish Government 2012a Scotland beats 2011 green energy target, March 29, http://www.scotland.gov.uk/News/Releases/2012/03/geenenergytargets29032012

Scottish Government 2012b Electricity Generation Policy Statement, The Scottish Government, Edinburgh, March, http://www.scotland.gov.uk/Topics/Business-Industry/Energy/EGPS2012/DraftEPGS2012

Sensfuß F, Ragwitz M and Genoese M 2007 Sustainability and Innovation, Fraunhofer Institute Working Paper S7/2007

SinoCast 2012 Interview with Liu Zhenya, SinoCast Daily Business Beat, May 5

Solar Portal 2012 If solar covered one percent of the UK it would meet the country's entire power demand, Solar Portal website, Oct 11, http://www.solarpowerportal.co.uk/news/if_solar_covered_one_percent_of_the_uk_it_would_meet_the_countrys_2356

Sovacool B 2008 Valuing the greenhouse emissions from nuclear power, *Energy Policy* **36** 2940–53, http://www.sciencedirect.com/science/article/pii/S0301421508001997

Sovacool B 2011 *Contesting the Future of Nuclear Power* (Singapore: World Scientific)

SRU 2011 Pathways towards a 100 % renewable electricity system, SRU, German Advisory Council on the Environment, Berlin, http://www.umweltrat.de/SharedDocs/Downloads/EN/02_Special_Reports/2011_10_Special_Report_Pathways_renewables.html

Szarka J, Cowell R, Ellis G, Stracham P and Warren C (eds) 2012 *Learning from Wind Power* (Basingstoke: Palgrave Macmillan)

TBL 2013 The Feasibility of Renewable Energy in Pakistan, Triple Bottom Line advocacy NGO http://www.tbl.com.pk/the-feasibility-of-renewable-energy-in-pakistan/

Teyssen J 2012 E.on confirms strategy focused on renewables *Windpower Monthly*, quoting E.ON executive Teyssen's comments to German business daily Handelsblatt, March 30, http://www.windpowermonthly.com/go/windalert/article/1124937/?DCMP=EMC-CONWindpowerWeekly

Tickell O 2012 Thorium: Not 'green', not 'viable', and not likely, NGO Briefing note, http://www.nuclearpledge.com/reports/thorium_briefing_2012.pdf

Timilsina G, Beghin J, van der Mensbrugghe D and Mevel S 2012 The impacts of biofuels targets on land-use change and food supply: A global CGE assessment *Agri. Eco.* **43** (3) 315–32, May, http://onlinelibrary.wiley.com/doi/10.1111/j.1574-0862.2012.00585.x/abstract

Toke D 2007 Renewable financial support systems and cost-effectiveness, *J. Clean. Prod.* **15** 280–7

Tzimas W 2011 Sustainable or Not? Impacts and Uncertainties of Low-Carbon Energy Technologies on Water, European Commission –Joint Research Centre, Institute for Energy, Power Point presentation to AAAS meeting, http://ec.europa.eu/dgs/jrc/downloads/jrc_aaas2011_energy_water_tzimas.pdf

UBA 2010 Energy target 2050: 100% renewable electricity supply, German Federal Environment Agency, Umweltbundesamt, Berlin, http://www.uba.de/uba-info-medien-e/3997.html

UCG 2013 Underground Coal Gasification Association, http://www.ucgassociation.org

UKAEA 2007 Fusion – a clean future, UK Atomic Energy Authority booklet

UN 2013 Sustainable Energy for All programme, United Nations, http://www.sustainableenergyforall.org/

UNEP 2008 Green Jobs, United Nations Environment Programme, http://www.unep.org/civil-society/Implementation/GreenJobs/tabid/104810/Default.aspx

UNDP 2013 Derisking Renewable Energy Investment, UN Development Programme, New York, http://www.undp.org/content/undp/en/home/librarypage/environment-energy/low_emission_climateresilientdevelopment/derisking-renewable-energy-investment/

UNFCCC 2012 Benefits Of The Clean Development Mechanism, United Nations Framework Convention on Climate Change report, Bonn

Vertrung 2013 Germany embassy translation – outline of energy plan, http://www.germany.info/Vertretung/usa/en/06__Foreign__Policy__State/02__Foreign__Policy/05__KeyPoints/ClimateEnergy__Key.html

Weinberg 2013 Weinberg Foundation, http://www.the-weinberg-foundation.org/

WFC 2013 From vision to action: A workshop report on 100% Renewable Energies in European Regions, World Future Council, Hamburg, http://www.worldfuturecouncil.org

Willis R and Wills J 2012 Co-operative renewable energy in the UK, Co-operatives UK, Manchester, http://www.uk.coop/sites/default/files/renewableenergy_0_0.pdf

Wind connection 2013 US Offshore grid proposal website, http://atlanticwindconnection.com

Wiser R and Bolinger M 2012, 2011 Wind Technologies Market Report, US Department of Energy, http://eetd.lbl.gov/ea/emp/reports/lbnl-5559e.pdf

WRG 2009 Charting Our Water Future, 2030 Water Resources Group, http://www.2030waterre-sourcesgroup.com/water_full/Charting_Our_Water_Future_Final.pdf

ZCA 2010 Zero Carbon Australia, Beyond Zero Emissions campaign, http://beyondzeroemissions.org/zero-carbon-australia-2020

IOP Publishing

Renewables
A review of sustainable energy supply options
David Elliott

Chapter 7

Afterword

On the other hand: sceptical views

I have suggested that the debate over energy futures is vital and should be engaged in widely. I therefore felt it might be helpful to have a brief overview of the structure and dynamics of the debate so far. This short account will inevitably be partial and contentious but, for what it is worth, I will try to map it out in rough terms, with apologies to social science actor network theorists, who no doubt could do better!

The starting point must surely be the policies and programmes of governments, local, national and international, augmented by the policies and plans of industrial groupings, which together set the scene for the debate. Indeed these organisations, and some of the individuals within them, are major players. Academics also play roles, but I want to focus on another set of players who also often enjoy media coverage: the lobby groups. Some are trade lobbies, some are local, but more usually national or international green lobby groups and environmental non-governmental organisations (NGOs) with, usually, clear policy agendas. In addition there are some quasi-independent 'think tanks' and special interest groups, usually with some sort of political policy agenda, overt or otherwise.

The activities and aims of the green NGOs are usually fairly transparent, but for some of the other lobby groups this may not quite be the case. What, for example, are we to make of the UK's Renewable Energy Foundation, which has a reputation for being very critical of wind energy? I am not suggesting anything sinister. However, in a healthy democracy, the policies, proposal and views of all agencies and groups need to be scrutinised and, if necessary, challenged, and the groups held to account. Green groups have no special claim to truth or credibility, but then neither do some of the other groups. They too need scrutiny, especially since, like the NGOs, they can be very influential.

Some organisations set themselves up to be ostensibly 'green' but arguably in practice may not be seen as such by most people, although 'green' is a very loose term. The UK's Country Guardian group, which is fiercely anti-wind power, would no doubt argue the point. Less credibly, so might some of the pro-nuclear groups, such as Supporters of Nuclear Energy (SONE) in the UK, but that is a matter of opinion. In common with celebrated UK 'Gaia' theorist James Lovelock and US climate expert James Hansen (Lovelock 2004, Kharecha and Hansen 2013), a handful of environmentalists have also backed nuclear power and have been much feted by the nuclear lobby (Brand 2009, Moore 2013, Lynas 2013). Some of the above have been critical of renewables, but here I will confine myself to some examples of major outputs from influential groups and individuals specifically on renewable energy.

John Etherington's 2009 book *The Wind Farm Scam—an Ecologist's Evaluation* is evidently seen as a definitive text by anti-wind groups. You do not have to read much of it to see why. He says that the wind power industry is determined 'to drive roadway after roadway through lonely places, to dump concrete in enormous quantity, to bulldoze acres of hillside into wind farms studded with gigantic, identically mass-produced steel and plastic monsters. This is akin to demolishing the great cathedrals for road stone or shredding the contents of the National Gallery to make wall insulation'. He adds 'as the developers have grabbed the remote lands of Britain, so their flailing blades perforce creep closer to habitations'. He describes wind turbines as 'wind monsters' spreading 'environmental harm' and sees anti-wind campaigners as 'heroic defenders of the land'.

This use of colourful language seems a little excessive. Some wind farms may have been poorly sited or planned, with not enough sensitive local consultation, but they can if necessary easily be removed and planning procedures improved. However, Etherington is unrepentant. Rather than 'twitching crucifixions of landscape', he recommends, as an alternative, nuclear power, which he claims 'could give secure supply of very large amounts of electricity'. He is also contrarian on the use of climate change as a justification for wind projects: 'It is not credible that the virtual-world output of the models can reliably be used to make policy decisions'.

So can this book be ignored as just a silly opinionated diatribe? No, since Etherington is an ecologist and academic of some standing, having been Reader in Ecology at the University of Wales, Cardiff, and a former editor of *The Ecologist* magazine. And the bulk of the book consists of a well written and detailed account of wind power, how it works and what problems there might be, with much of this being respectably done, even if there are occasional lapses and errors. Some of the errors are technical. He is not an engineer and occasionally slips up on details, some of which are important. But it is unremittingly negative. Professor John Twidell provided a detailed and well measured critique, pointing out the errors and misunderstandings, in a review in the academic journal *Wind Engineering* (Twidell 2010). He felt that underlying Etherington's technical and economic arguments was a fundamental dislike of the technology and its presence in the landscape. That is a valid view, if rather aggressively purveyed.

The more professionally presented critical views on wind power's economics from UK-based groups like the Renewable Energy Foundation and the Global Warming Policy Foundation have also been met with responses from academics and others, for example in relation to the report 'Why wind power is so expensive' by Professor Gordon Hughes from the University of Edinburgh, produced for the Global Warming Policy Foundation (Hughes 2012a). He claimed that wind power's 'intermittency' was a major problem.

As he summarised it in a subsequent response to the media debate that ensued: 'to maintain secure reserve margins, each MW of wind generating capacity has to be backed by approximately 1 MW of generating plant which can be run on demand. There is absolutely no saving in CO_2 emissions because the gas plants carry on running as before but they are just feeding less electricity into the grid' (Hughes 2012b).

This seems odd. When not delivering full power, gas-fired power plants use less fuel, so produce less CO_2. Moreover, they do not have to back up each individual wind turbine.

The variations in output from individual turbines, and the often differing variations from others at other sites, as well as the variations from other types of power plant, will be aggregated by the grid. It is the end result that has to be balanced. In addition, there are many other grid-balancing options. So combined cycle gas turbine (CCGT) plants do not have to take all the strain. And in any case, as I noted earlier, there are new, more flexible CCGTs emerging which can ramp up and down rapidly with much less efficiency loss than existing designs.

The basic point on emissions and 'intermittency' was made elegantly in 2008, in a report from the House of Lords Select Committee on Economic Affairs on 'The Economics of Renewable Energy'. Referring to an argument made to the Committee by a witness from the Renewable Energy Foundation, who claimed that any carbon savings from wind power were offset by the need to run conventional fossil fuel plants at part load to balance the fluctuations in wind output, the Committee concluded that 'the need to part-load conventional plant to balance the fluctuations in wind output does not have a significant impact on the net carbon savings from wind generation'.

The debate over wind 'intermittency' will nevertheless no doubt continue. For an extensive, authoritative report on the issue see the IEA's 'Harnessing Variable Renewables' (IEA 2012). Technical and economic arguments of this sort may of course not satisfy those who are fundamentally opposed to wind farms, or who prefer other options; it is the same for those who object fundamentally to nuclear power. In the end, attitudes to technology, as to anything else, may come down to beliefs and values, hopes and fears. They may not always be rational, in reductionist scientific terms, but they may reflect important human concerns not always captured by 'objective' assessments.

This is not the place to explore how attitudes and beliefs are formed or changed, but they do clearly shape the policy positions adopted by various groups, often reflecting political orientations. Those on the left tend to support change; those on the right tend to be conservative. While most green NGOs tend to have generally progressive values, there has certainly been no shortage of negative contentions in relation to wind and some other renewables from various right-of-centre think tanks, including, in the UK, the Centre for Policy Studies, the Institute of Economic Affairs, the Adam Smith Institute, and Civitas.

The report from the last group was entitled 'Electricity Costs: The Folly of Wind Power'. Similar, if more aggressively expressed, views have also emerged from anti-wind groups like the European Platform Against Windfarms and its US counterparts (Anti-wind platforms 2013), including the at least technically credible Energy Collective group (Energy Collective 2013). Australia's Brave New Climate website can also be useful, despite its very pro-nuclear stance (Brave New Climate 2013).

Definitely not pro-nuclear, but very critical of the view that renewables could supply all of humankind's needs, Australian academic Ted Trainer might also be seen as a 'contrarian', but one coming from a radical 'deep green' viewpoint. He argues that renewables cannot sustain energy-intensive societies based on economic growth and looks to a massive reduction in energy use and radical transformation of consumerist culture. This is a valid, if challenging, political prescription. Certainly, as I have argued, technically, the more we can reduce demand, the easier it would be to meet energy needs with renewables. However, that is not his view. Instead, some of Trainer's criticisms seem based on pessimism about the technology, e.g. he sees wind as being limited to a 25% global input. He also has doubts about how much can be obtained from concentrated solar, as well as concerns about the capital costs of a renewable future (Trainer 2010).

The various very positive '100% renewables globally by 2050' assessments that have emerged recently give a very different view of the potential and costs of renewables. Some may be too optimistic. That was what Trainer argued in a critique of Delucchi and Jacobson's '100% renewables' paper in *Energy Policy*, claiming that they had not dealt sufficiently with intermittency and costs (Trainer 2012), although these authors came back with a fairly convincing rebuttal. Trainer had argued that renewable energy could supply the world only if the world 'embraces frugal lifestyles, small and highly self-sufficient local economies, and participatory and co-operative ways in an overall economy that is not driven by growth or market forces'. Delucchi and Jacobson say 'This vision may or may not be desirable, but it was found in our study not to be necessary in order to power the world economically with wind, water, and solar energy' (Delucchi and Jacobson 2012).

Trainer may be right to warn us not to overstate what renewables can do. However, he may also risk undermining them. Indeed, it is almost as if he does not want renewables to work, so we have to get on with the more important social changes. Certainly some see radical social and lifestyle changes as vital and as part of an urgent political and economic process of change. But most would include renewables as a central part of that transition (Abramsky 2010).

Clearly there is plenty of room for debate and, to an extent, the various contrarians, from their various political viewpoints, make a contribution, even if it is not always a welcome one. As I have indicated, there have been strong rebuttals of some contrarian views, but there is a healthy, if at times rather bilious, debate going on, much of it on the internet. Some of it is technical, some of it political. If you feel so inclined, you can join in. It is good to be exposed to divergent views, and like the Energy Collective and Brave New Climate, the Renewable Energy Foundation (REF 2013) certainly has access to technical expertise and data. You may, however, find it more productive to focus on other sources for analysis. I leave you to judge where to find those, although personally, for all its faults, I find the UK Claverton Energy Group's e-group practical engineering-orientated discussions enlivening and informative, if at times chaotic!

Finally, can I point out that, while it is vital to get the facts straight wherever possible, and be aware that developers, promoters and enthusiasts can overstate their case, there will always be some uncertainty, disagreement, prejudices and odd distortions. For example, those who are confident about the prospects for developing high-temperature liquid sodium-cooled fast breeders, or molten fluoride salt thorium reactors, may sometimes, a little oddly, baulk at the technical difficulties they see as being associated with what are surely relatively much less complex wind, wave and tidal technology. More generally, it is possible to run energy scenarios with very different outcomes, based on differing assumptions about what is technically and economically credible. There is an element of judgement involved (Pugwash 2013).

Objectivity, while very desirable, is sometimes quite hard to sustain, and indeed may not always be possible, especially when thinking about future systems and developments.

We may still need faith and hope, even in matters of technology. That links to a current contention in policy circles that care has to be taken to avoid 'optimism bias'. On balance, I think I would prefer to avoid 'pessimism bias'.

For example, as REN21 has pointed out, in 2000 the International Energy Agency projected 34 GW of wind power globally by 2010, while the actual level reached was 200 GW. The World Bank in 1996 projected 9 GW of wind power and 0.5 GW of solar PV in China by 2020, while the actual levels reached in 2011, nine years early, were 62 GW of wind power and 3 GW of solar PV. Looking to the future, in 2012 REN21 interviewed 170 energy experts, and found that most industry experts believed that the world could reach at least 30–50% shares of renewables, while some experts advocated 100% or near-100% futures (REN 21 2013). I tend to support the latter view.

That may prove to be unrealistic. However, although you should remember that my likely biases may need to be taken into account, and you should immerse yourself in all sides of the debate, I close with Bertrand Russell's dictum that 'Science may set limits to knowledge, but should not set limits to imagination'.

Summary points

- The debate over energy is important, but it is sometimes arguably disfigured by lobbyist's agendas and biases.
- It is important to scrutinise and carefully assess all views and assertions from all sides.
- Although objectivity is desirable, vision, faith and hope may also be important.

References

Abramsky K (ed) 2010 *Sparking a World-wide Energy Revolution* (Oakland: AK Press)

Anti-wind platforms 2013 European Platform against windfarms, http://www.epaw.org/ North American Platform against windfarms, http://www.na-paw.org

Brand S 2009 *Whole Earth Discipline* (New York: Viking Penguin)

Brave New Climate 2013 Australian Energy Issues website, http://bravenewclimate.com

Claverton Energy Group 2013 UK-based energy practitioner's network, Bath, http://www.claverton-energy.com/

Delucchi M and Jacobson M 2012 Response to 'A critique of Jacobson and Delucchi's proposals for a world renewable energy supply' *Energy Policy* **44** 482–4, http://www.sciencedirect.com/science/article/pii/S0301421511008731

For Trainers reaction see http://www.sciencedirect.com/science/article/pii/S0301421512008658

Energy Collective 2013 Energy Collective website, Vermont, http://theenergycollective.com

Etherington J 2009 *The Wind Farm Scam—an ecologist's evaluation* (London: Stacey International)

Hughes G 2012a Why wind power is so expensive, Global Warming Policy Foundation report, http://www.templar.co.uk/downloads/hughes-windpower.pdf

Hughes 2012b Global Warming Policy Foundation website, http://www.thegwpf.org/gordon-hughes-response-to-goodall-lynas/

IEA 2012 Harnessing Variable Renewables, International Energy Agency, http://www.oecdbookshop.org/oecd/display.asp?sf1=identifiers&st1=612011171P1&LANG=EN

Kharecha P and Hansen J 2013 Prevented mortality and greenhouse gas emissions from historical and projected nuclear power *Environ. Sci. Technol.*, March 15, http://pubs.acs.org/doi/abs/10.1021/es3051197?journalCode=esthag

Lovelock J 2004 James Lovelock: Nuclear power is the only green solution *The Independent*, May 24, http://www.independent.co.uk/voices/commentators/james-lovelock-nuclear-power-is-the-only-green-solution-6169341.html

Lynas M 2013 Website run by Mark Lynas, http://www.marklynas.org/

McCaffery M 2012 Busting the myths (again) on wind energy *Business Green*, Aug 17, http://www.businessgreen.com/bg/opinion/2199358/busting-the-myths-again-on-wind-energy

Moore P 2013 Clean and Safe Energy US nuclear lobby group, http://cleansafeenergy.org/

Pugwash 2013 Pathways to 2050: Three possible UK energy strategies, British Pugwash, London, http://www.britishpugwash.org/recent_pubs.htm

REF 2013 UK Renewable Energy Foundation, http://www.ref.org.uk/

REN 21 2013 Renewables: Global Futures Report, Renewable Energy Policy Network for the 21 Century, Paris, http://www.ren21.net/gfr

Trainer T 2010 Can renewables etc. solve the greenhouse problem? The negative case *Energy Policy* **38** (8) 4107–14, August, http://www.sciencedirect.com/science/article/pii/S0301421510002004 For a summary see http://www.countercurrents.org/trainer090710.htm

Trainer T 2012 A critique of Jacobson and Delucchi's proposals for a world renewable energy supply *Energy Policy* **44** 476–81, May, http://www.sciencedirect.com/science/article/pii/S0301421511007269

Twidell J 2010 Book review of Etherington's The Wind Farm Scam *Wind Engineering* **34** (3) 335–50

Lightning Source UK Ltd.
Milton Keynes UK
UKOW06n0640091014

239819UK00001B/5/P

9 780750 310413